DYNAMICS AND DIVERSITY

SOIL FERTILITY AND FARMING LIVELIHOODS IN AFRICA

Case Studies from Ethiopia, Mali and Zimbabwe

Edited by
Ian Scoones

Earthscan Publications Ltd, L

First published in the UK and USA in 2001 by
Earthscan Publications Ltd

ISBN: 1 85383 820 9 paperback
1 85383 819 5 hardback

Typesetting by PCS Mapping & DTP, Newcastle upon Tyne
Printed and bound by Creative Print and Design, Ebbw Vale
Cover design by Danny Gillespie
Cover photos reproduced by kind permission of: CGIAR (Consultative Group on International Agricultural Research), The World Bank; CIMMYT (Centro Internacional de Mejoramiento de Maiz y Trigo); ICRAF (International Centre for Research in Agroforestry), Nairobi, Kenya; Chris Reij

For a full list of publications please contact:
Earthscan Publications Ltd
120 Pentonville Road
London, N1 9JN, UK
Tel: +44 (0)20 7278 0433
Fax: +44 (0)20 7278 1142
Email: earthinfo@earthscan.co.uk
http://www.earthscan.co.uk

22883 Quicksilver Drive, Sterling, VA 20166–2012, USA

Earthscan is an editorially independent subsidiary of Kogan Page Ltd and publishes in association with WWF-UK and the International Institute for Environment and Development

A catalogue record for this book is available from the British Library

Library of Congress Cataloging-in-Publication Data

Dynamics and diversity : soil fertility management and farming livelihoods in Africa : case studies from Ethiopia, Mali, and Zimbawe / edited by Ian Scoones.
p. cm.
Includes bibliographical references (p.).
ISBN 1-85383-820-9 (paper : alk. paper) – ISBN 1-85383-819-5 (cloth : alk. paper)
1. Soil fertility–Ethiopia–Management. 2. Soil fertility–Mali–Management. 3. Soil fertility–Zimbabwe–Management. I Scoones, Ian.

S599.5.A1 D96 2001
631.4’22’096–dc21

2001004209

This book is printed on elemental chlorine-free paper

CONTENTS

List of Tables

List of Figures

List of Boxes

Foreword

The reasons for the very significant gap between potential and realized food production in sub-Saharan Africa are multiple and complex. The decline in fertility observed for many areas of soil has been described as the single most important factor. Although this is a challengeable statement it undoubtedly refers to an ever-present reality for the majority of farmers in the continent – that optimizing the nutrient balance on their farms is one of the most difficult of the many agricultural management challenges they face.

A central feature of this book is the documentation of the great variety of ways in which farmers have dealt with this problem. More importantly it also gives excellent insight into the ways in which the soil fertility issue interacts with a multiplicity of other factors which impact on farm production – biological, economic, social and political. Scientists, with their strong disciplinary adherences, apply the power of reductive research to these issues and often provide solutions which are valid within their own limits, but which are difficult to apply because of the lack of attention to these interactive factors.

The work reported in this book helps to resolve this disjunction between formal scientific method and the realities of farm management. Scientific methods of varying degrees of formality are used to document and analyse the soil fertility 'problem', the factors which influence it and farmers' coping strategies. The replication of this across different countries, environments and communities permits the drawing of commonalities as well as distinctions. The major benefit that may be gained from this is to inform scientists – not just with data but with insights into the realities of the totality of the farming enterprise. The challenge is then to identify those 'entry points' where formal scientific knowledge can be employed to enhance the system as a whole. A strong case can indeed be made that soil fertility management is a very significant entry point because of the many interactions it has with other components and because of the long-term nature of the effects that result from changes in soil nutrient status.

This book is thus to be recommended not just for the information and insights it provides with respect to the specific issue of soil fertility management, but also because of the major questions it provokes about the application of scientific research to the challenges of sustainable agriculture under the prevailing conditions in African countries.

Professor Mike Swift
Director, Tropical Soil Biology and Fertility Programme, Nairobi

PREFACE

Soils are critical to agriculture and in turn to food security and livelihoods. With a large proportion of the population in Africa dependent on small-scale agriculture, the sustainable management of the soil resource is a high priority issue. This is increasingly recognized in national and international policy debates. Yet such debates are often informed by limited insights into the immense diversity and complex dynamics of real farming settings. Too often a picture of crisis and collapse informs policy statements, suggesting the need for a particular type of intervention and management.

This book aims to look behind such statements by asking searching questions about what is really going on. Through the detailed analysis of case studies from Ethiopia, Mali and Zimbabwe a much more nuanced picture is built up. In some places, for some people, soils are improving and sustainable land management options are being encouraged. In other situations a more negative scenario holds, where soil degradation potentially threatens the long-term viability of agriculture. The practices of soil management are seen to be intimately bound up with people's broader livelihood strategies, with a whole complex of factors impinging on the success or otherwise of sustainable soil management. Ecological dynamics, socio-cultural factors, institutional arrangements and policies of various sorts all have an impact.

Such dynamics and diversity require an interdisciplinary approach to analysis, linking field-level practice to policy debates at national and international scales. This book is based on research carried out by teams of researchers from Africa and Europe over three years in a range of contrasting locations. Natural science investigations of soil properties and nutrient flow dynamics were linked to social science analyses of social difference, institutions and policy, set within an understanding of the historical context. Together, such analyses informed a process of action research with farmers and researchers working together on practical solutions in the field.

The research results add up to a new approach to looking at soil management issues in Africa, with significant implications for development policy and practice. An interdisciplinary methodology, for example, moves us away from the often simplistic, aggregate technical diagnoses that have informed many policy statements to date. Understanding soils in the context of livelihood systems also suggests new ways of thinking and acting. Overall, the results suggest a more positive view of the prospects for sustainable agriculture in small-scale farming systems in Africa, with a fundamental challenge to the overridingly negative views of crisis and collapse which have dominated the policy debate. But this does not mean that all is well. The research also points to the critical need to develop new technologies and management practices which are suited to the diversity of farmer needs and settings. It also points to the need to take seriously institutional and policy issues, across a variety of scales, when addressing the challenges of natural resource management in Africa.

The research reported in this book has involved a lot of people. The research teams (see details in Chapters 2 to 4) involved 38 researchers in a variety of different capacities, ranging from field data collection to research coordination. In Ethiopia the NGOs FARM-Africa and SOS Sahel provided the institutional base for the project, while in Mali the Institut d'Economie Rurale's Niono team led the work. In Zimbabwe the Farming Systems Research Unit of the Department of Research and Specialist Services in the Ministry of Agriculture was the coordinator. The commitment of the respective organizations and the staff involved in the research has been critical to the success of the work. In addition, farmers in the research sites, together with extension workers, local government officials and others, have contributed considerable amounts of their own time in collecting data, as well as discussing and analysing the research findings. Without such inputs, and especially the continuous, considered critique and reflection from the field level, the grounded picture of real farming settings which the book aims to capture would not have come through.

The overall research programme was coordinated by the Drylands Programme at the International Institute for Environment and Development, and considerable thanks are due to Camilla Toulmin and her team in Edinburgh and London. Coordination was shared, particularly for support to work in Ethiopia and Zimbabwe, by the Environment Group at the Institute of Development Studies, University of Sussex. Partners based at the Royal Dutch Tropical Institute in The Netherlands have also been key in the research, providing vital support work, particularly in Mali. The research group has met on a number of occasions during the research process to share ideas, revise plans and reflect on findings. The meetings in Tanzania, Ethiopia, Zimbabwe and Mali have been vital in fleshing out the core themes that make up this book. We have been helped in this process, particularly during the early stages, by very productive interactions with a parallel project being coordinated by the Tropical Soil-fertility and Biology Programme based in Nairobi. The research, of course, would not have been possible without the financial support of the European Union Science and Technology for Development Programme (grant number: TS3-CT94–0329). We are most grateful to Mario Catizzone and Dirk Pottier for their support and encouragement.

This book has been compiled and edited by Ian Scoones on the basis of a wide range of reports and project outputs. Editorial assistance from Camilla Toulmin and Annette Sinclair has been invaluable. The aim has been to produce a synthetic product reflecting the richness of the case studies, drawing lessons for development policy and practice more broadly. We hope the book will both provoke further debate and be of interest to a wide audience committed to environment and development issues in Africa.

Ian Scoones
Institute of Development Studies
Brighton
May 2001

List of Acronyms and Abbreviations

ABLH	Association for Better Land Husbandry
AN	ammonium nitrate
AVs	Associations Villageoises
CFDT	Compagnie Française pour le Développement des Textiles
CMDT	Compagnie Malienne pour le Développement des Fibres Textiles
DAP	di-ammonium phosphate
DC	District Commissioner
DRSS	Department of Research and Specialist Services
ESAP	Economic Structural Adjustment Programme
FAO	Food and Agriculture Organization (UN)
FSRU	Farming Systems Research Unit
ha	hectare
IER	Institut d'Economie Rurale
ILEIA	Centre for Research and Information on Low-External-Input and Sustainable Agriculture
IMF	International Monetary Fund
IPCC	Intergovernmental Panel on Climate Change
ISFM	integrated soil-fertility management
masl	metres above sea level
MPP	Minimum Package Programme
NC	native commissioner
NEAP	National Environmental Action Plan
NGO	non-governmental organization
PA	peasant association
PADETES	Participatory Demonstration and Training Extension Programme
PNVA	Programme National de Vulgarisation Agricole
RFM	resource-flow map
SARDC	South African Research and Documentation Centre
SFI	Soil-Fertility Initiative
SOM	soil organic matter
TSBF	Tropical Soil Biology and Fertility programme (Nairobi)
UNEP	United Nations Environment Programme
WADU	Wolamu Agricultural Development Unit

Chapter 1

Transforming Soils: The Dynamics of Soil-fertility Management in Africa

Ian Scoones

Introduction

Issues of soil management are at the top of the international policy agenda for Africa these days. Many statistics are marshalled to support the view that something must be done about declining soil-fertility and increasing soil degradation. If measures are not taken, it is argued, there will be a continuing 'downward spiral' of increasing land degradation and rural poverty. Investment in agriculture, and particularly in soils and their management, must be a high priority for public funding if Africa is to achieve any level of agricultural success in its struggle for development.

While this summary of the mainstream position is in some senses a caricature, it does resonate with many of the statements from international agencies of recent years, as will be shown below. This refrain of concern, of course, is not a new one, and the history of intervention in soils management in Africa has been fuelled by such calls to action based on dramatic predictions about future collapse.

One of the main messages of this book is that we must be extremely wary about such generalized statements. The real world of farmers, explored in detail with three country case studies in subsequent chapters, is much more complex. Issues of spatial and temporal dynamics, of diversity and difference, of history and change, of socio-economic setting and relationships, of policy context and trends are central to a more balanced analysis of what is happening. While such detailed perspectives incorporate elements of the mainstream position on soil-fertility change, they also point to new insights and new directions for intervention and policy.

An alternative conceptual basis for understanding soils and their management can be derived from different disciplinary interactions and combining methodological tools from which to suggest new directions for the soil management debate in Africa. Such new directions contest the simplistic statements generated by aggregate statistics and undifferentiated analysis, and provide a more comprehensive understanding of local complexity, diversity and dynamics. Such insights demonstrate why policy and intervention need to be more rooted in local settings and local understandings (see Chapter 6).

Drawing on a comparative review of the case studies from Ethiopia, Mali and Zimbabwe presented in Chapters 2 to 4, this overview synthesizes some of these new directions. Following a look at how the soil-fertility debate in Africa is conventionally understood, the case studies are introduced, highlighting both contrasting and similiar features across sites. Next, the type of evidence for soil-fertility change in Africa is reviewed, with an historical look at how scientists have understood the issue. The range of 'narratives' which have informed mainstream policy thinking over time is identified, along with their underlying theoretical assumptions and methodological commitments. In the following section, an alternative perspective is outlined, which attempts to take the spatial and temporal variability of changes in soils into account. A conceptual framework centred on an understanding of diversity and dynamics is offered, together with some reflections on the methodological implications of such an approach. By taking examples from the case studies, the implications of interpreting soil change processes with such an alternative lens are explored. The broader implications for research–action approaches at field and policy levels are, in turn, further explored in Chapters 5 and 6. In the final chapter, we turn to an examination of the range of determinants, both endogenous and exogenous, of the multiple pathways of agricultural and environmental change evident across the case study sites, setting the analysis of soil-fertility change within a broader livelihoods context.

The current policy debate[1]

The current policy debate on soil management and agricultural development in Africa is characterized by a strong storyline describing the nature and scale of the problem, its causes, its consequences and the intervention options available for doing something about it. Such policy 'narratives' (Roe, 1991; Leach and Mearns, 1996) suggest a story about what the problem is and what should be done about it. While there are, of course, variations, with different emphases and nuances, the basic argument and, importantly, the conclusions remain broadly the same across a wide range of sources.[2] With concerns about 'desertification' raised by international debates on the future of the African environment, such themes have gained great prominence in many quarters.[3] An essentially negative picture is painted of a 'downward spiral' within which increasing environmental degradation is associated with growing poverty, a situation that requires major investments in soil-fertility management at national and continental scales.

Today, soil-fertility decline – and particularly what has been termed 'nutrient mining' – is seen to be widespread in sub-Saharan Africa, linked especially with population increase. Declining yields, as a result of continuous cropping on exhausted soils, are shown to be a threat to food and livelihood security across the continent. The major challenge therefore is to reverse the tide of nutrient loss and increase the soil stocks through recapitalization initiatives. In lauching the Africa-wide Soil-fertility Initiative, the World Bank and Food and Agriculture Organization (FAO) (1996, p1) argue that:

> *The factor which impedes agricultural growth the most fundamentally is continuous mining of soil nutrients throughout Africa... Without restoration of soil-fertility, Africa faces the prospects of serious food imbalances and widespread malnutrition and likelihood of eventual famine.*

Similarly, ICRAF scientists (Buresh et al, 1997, pxi) argue:

> *Sub-Saharan Africa is the last continent facing massive problems of food security because of decreasing per-capita food production. Extreme poverty, widespread malnutrition and massive environmental degradation are direct consequences of a policy environment that results in large-scale nutrient mining.*

The underlying causes of such degradation are seen to be associated with the combination of population growth, poverty and poor agricultural practices. The neo-Malthusian 'nexus' argument put forward by the World Bank identifies a 'downward spiral' of increasing low productivity and land degradation (see Cleaver and Schreiber, 1995). For example, the World Bank/FAO concept paper argues that, in many parts of sub-Saharan Africa:

> *The nexus of rapid population growth and high population densities, low productive agriculture, and depletion of natural resources has created negative synergies that exacerbate existing conditions of soil nutrient mining and underdevelopment, thus creating a vicious circle of poverty and food insecurity* (World Bank/FAO, 1996, p4).

The result is seen to be a cycle of poverty and vulnerability linked to continued resource degradation:

> *In regions with fallow farming or integrated livestock farming... growing population pressure compels farmers to replant fallow land before soil-fertility has been restored or to work marginal land only suitable for pasture or forestry. The outcome is a downward spiral of instability–unsustainability. The spiral ends in a vicious circle of 'low input–low yield–low income'* (Steiner, 1996, p13).

This is certainly rather a pessimistic and depressing story, one that is repeated in the context of many of the dominant policy commentaries in Ethiopia, Mali and Zimbabwe. Contemporary national policies and donor strategies in each of the countries focus on the potential negative consequences of increasing population and heightened risks of environmental degradation leading to threats to agricultural production and rural livelihoods. For example, Gakou et al (1996, pi) comment on the Malian situation:

> *In Mali, as in most Sahelian countries, the constraints of climate, demographic pressure and the irrational exploitation of natural resources causes degradation of agricultural land… This degradation is often aggravated by the fragility of cultivated lands and the poor adaptation of production systems and techniques.*

Similarly in Ethiopia, the Soil-fertility Initiative concept paper (Wales and Le Breton, 1998, p6) notes:

> *The mechanisms promoting soil degradation in Ethiopia are much the same as elsewhere in Africa. Forest clearance and soil exposure, poor crop cultivation practices including cultivation on steep slopes, removal of crop residues and the burning of dung, and overgrazing, all contribute to soil loss. Indirect causes include poverty, insecure land tenure, population growth and economic policies which do not encourage good husbandry of land resources.*

Given these dominant positions on policy it is necessary to ask: what is the evidence for this rather gloomy, pessimistic position? Is the situation so universally doom-laden, or is there evidence for a more optimistic view? Are there alternative – or at least more nuanced – perspectives, based on different methods and interpretations? Do these, in turn, suggest different strategies for what to do and how to do it? These are the questions which subsequent sections of this chapter, and the case study chapters that follow, will examine.

UNDERSTANDING SOILS IN AFRICA

So what is the knowledge base upon which current researchers, planners and policy makers draw? From the early colonial era scientists have invested considerable efforts in trying to understand Africa's soils.[4] Coming from temperate regions, colonial scientists were intrigued by the ancient, heavily weathered soil formations, the rapidity of the mineralization and decomposition processes, and the spectacular nature of soil erosion, particularly gullies. A set of views about African soils emerged which continues to inform scientific perceptions. These included beliefs that African soils are inherently infertile, that erosion is a major issue, and that substantial amounts of soil-regenerating materials must be added to ensure successful production. These perspectives on tropical soil science are only partly correct (see Greenland et al, 1992). Many soils, particularly those derived from more recent volcanic activity, are highly fertile (Sanchez and Logan, 1992); soil processes vary considerably between different soil types, temperatures and moisture regimes (Woomer and Swift, 1994); and gullies, while impressive, may not be the most important soil degradation issue (Stocking, 1994).

As part of the process of colonial occupation, mapping and survey teams were sent out to document the new territories. These usually included a significant soil survey component.[5] Classification and mapping had long been an important component of soil science, dating back to the earliest attempts in Germany in 1862 (Russell, 1988). The earliest soil map of Africa was produced in 1923 (Shantz and Marbut, 1923), but this was highly schematic. It was not until soil surveyors undertook regional studies that a more detailed understanding of soils emerged.[6]

The study of soil erosion was one of the early preoccupations of scientists and technicians. This had its origins in the late 1920s when Haylett established run-off plots at the University of Pretoria in South Africa (Hudson, 1971). Similar plot-based experiments were established in various countries during the 1930s and 1940s when policy concern about soil erosion was reaching a peak (Tempany et al, 1944; Tempany, 1949). At this time, commentators predicted soil erosion would lead rapidly to the complete collapse of farming if protection was not afforded to the land (eg Lowdermilk, 1935). The result was increased investment in soil erosion prevention measures across the continent, and further research into this issue.

Soil-fertility maintenance was another theme which attracted the attention of colonial scientists from the early part of the century. In particular, concern was raised about the longer term prospects of monocropping. The result was the establishment of trials to look at different rotational systems accompanied by a variety of input strategies (Greenland, 1994; Swift et al, 1994; Bekunda et al, 1997; Pieri, 1995).[7] Up to the 1950s, the majority of recommendations focused on combining legume-based rotations with organic-based inputs such as cattle or green manure, composted in a variety of ways, possibly with the addition of minerals such as rock phosphate or lime, depending on the conditions (Watts Padwick, 1983). From the 1950s, however, a growing emphasis on inorganic mineral fertilizers can be seen.[8] This resulted in the elaboration of numerous yield–response curves, under a wide range of settings, resulting in the development of fertilizer recommendations and packages for most countries.

However, due to changing economic circumstances and growing concerns about environment and health, a more recent shift can be detected which emphasizes a more integrated soil-fertility management approach.[9] Today, research efforts encompass a far wider range of technical issues, ranging from legume innoculation technologies to agroforestry.[10] Many of the concerns of the 1930s with green manuring, composting and manure management have returned to the top of scientists' research agendas.[11] The integrated soil-fertility management approach is supported by work which looks at the interaction of biological, physical and chemical processes in the soil, and which emphasizes the need to understand soil processes in order to increase the efficiencies of use of different nutrient inputs (Woomer and Swift, 1994; Woomer and Muchena, 1996; Cadisch and Giller, 1997).

Emerging conclusions

A number of broad conclusions can be identified which emerge from these various fields of research on African soils.[12] Experimental and survey work has described the range of soils found in Africa in some detail, highlighting, in particular, where the major macronutrients are limiting. Large areas of the continent with old and weathered soils are severely deficient in the major nutrients; in other areas the nutrient content of soils may be high, but this may not be available for use by plants due to immobilization and fixation (Buresh and Smithson, 1997; Warren, 1992). Research also demonstrates how limiting factors interact, both within a single time period and over time. Under different conditions in the same soil, either nitrogen, phosphorous, water or micronutrients may be the key limiting factors. Work which links an understanding of soils with plant growth and physiology also highlights the many points at which a certain factor may limit plant growth. Increasing the efficiency of nutrient use may therefore require attention being paid to the interacting effects on crop yields of uptake, utilization, replenishment and application efficiencies (Noordwijk, 1999). Such work emphasizes the importance of increasing plant growth potential not simply through the addition of external inputs, but also through increasing the efficiencies of nutrient use by careful attention to the placement and timing of input applications (Woomer et al, 1994).

Longer-term experiments show that when cultivation starts, yields decline rapidly (over three to four years) to a low-level equilibrium (Syers, 1997). Declines in soil organic matter (SOM) are found to be particularly significant, with 5 per cent losses on sands and 2 per cent losses on heavier soils being recorded each year (Pieri, 1995). A threshold effect has been observed; if SOM is reduced to around 1 per cent the response to fertility inputs is significantly reduced (Lal, 1995). Long-term soil amendment experiments show that yields can be boosted above the low-level equilbrium amount, but this is not sustained, particularly for inorganic fertilizers; this is less true for organic inputs and systems in which rotation is a key component. Overall, the best and most sustained long-term response to soil amendments is found where organic and inorganic sources are mixed (Swift et al, 1994).

Soil erosion has been documented in all parts of the continent, with plot-level losses ranging between 0.1 and 138 tonnes per hectare per year (t/ha/year) (Scoones and Toulmin, 1999). Soil loss from arable plots of around 40–50t/ha/year is quite typical (Lal, 1984; 1995). Clearly soil erosivity increases with steeper slopes, more rainfall, and less ground cover. However, the total soil losses at a catchment level are much less than would be suggested by the plot measures. Catchment studies are few and far between, but all show that processes of soil redistribution and deposition are important. A similar conclusion can be drawn from studies that look at siltation levels. While siltation of dams and other water bodies is a significant problem, the amount of soil deposited is only a relatively small proportion of the total lost in the landscape (see Walling, 1984 for Zimbabwe), as much of the soil is redistributed rather than permanently lost from productive use.

In recent years nutrient balance studies across Africa, pitched at a range of scales, have looked at how inputs and outputs match up in terms of key nutri-

Table 1.1 *A summary of nutrient balance studies in Africa*[13]

Scale	*Site*	*Rainfall mm/yr*	*Unit*	*Balance kg/ha/year* *Nitrogen*	*Phosphorous*	*Source*
Continental	Sub-Saharan Africa			–22.0	–2.5	Stoorvogel et al (1993)
Country	Mozambique		Smallholder rainfed			Folmer et al (1998)
			Cassava	–48.0	–9.0	
			Maize	–48.0	–10.0	
Region	South-western Kenya	1350–2059	Kisii district	–112.0	–3.0	Smaling et al (1993)
	Southern Mali		Region	–25.0	0.0	van der Pol (1992)
			Maize	–29.0	0.0	
			Millet	–47.0	–3.0	
			Fallow	–5.0	0.5	
	Southern Mali	700–1200	Production system			Breman et al (1990)
			'Average'	–13.0	–	
			'Intensive'	–21.0	–	
Village/site	Eastern Madagascar	2000–3500	Long-term shifting cultivation			Brand and Pfund (1998)
			Site	–30.0	–0.4	
			Catchment	–12.0	–0.2	
	Uganda	1050–1300	Farm land			Wortmann and Kaizzi (1998)
			Site 1	–208.0	–80.0	
			Sites 2	–67.0	–9.0	
	Burkina Faso (Sahelian zone)	450	Village field			Krogh (1995)
			Sandy	0.1	0.4	
			Loamy	–5.6	–0.3	
			Clay	–9.9	–0.2	
Farm	Western highlands, Kenya	1600–1800	Farm (inc hedgerows)	–86.0	–3.8	Shepherd et al (1995); Shepherd and Soule (1988)
	Kisii, Kenya	1200–2100	Farm	–102.0	–2.0	van den Bosch et al (1998); de Jager et al (1998)
	Kakamega, Kenya	1650–1800		–72.0	–4.0	
	Embu, Kenya	640–2000		–55.0	9.0	
	Southern Ethiopia		Field			Eyasu et al (1998)
	Upland	1250	Homefield	–3.0 to –4.5	4.0 to 8.0	
			Outfield	–54.0 to –95.0	3.0 to 6.5	
	Lowland	800	Homefield	–4.0 to –24.0	3.0 to 10.5	
			Outfield	–20.0 to –40.5	–1.0 to 6.5	
	Bukoba district, Tanzania	1000–2100	Banana homefields			Baijukija and de Steenhuijsen Piters (1998)
			High rainfall	–76.0		
			Without cattle	80.0	–5.0	
			Zero grazing		42.0	
			Low rainfall			
			Without cattle	–49.0	–1.7	
			Zero grazing	31.0	23.5	

West Tanzania	800–950	Field			Budelman et al (1995)
		Sandy (cotton/ cassava)	–17.0	0.0	
		Loamy/clay (rice)	–56.0	–7.0	
North-east Nigeria	820	Farm	–28.2 to 2.5	–3.4 to 2.9	Harris (1996)
North-east Nigeria	360	Farm	–8.98 to 1.18	–0.81 to 1.5	Harris (1997)
Southern Mali	800–900	Farm	34.4	5.4	Defoer (1998)
		Field	–10.9	–14.1	

ents. Table 1.1 offers a compilation of such studies carried out over recent years at different scales and from different parts of Africa. These data show a consistent pattern of negative balances for nitrogen. Phosphorous balances show a more mixed story, with some cases of accumulation. Balances are more negative in the higher-rainfall, more productive sites (due to increased erosion, more harvest removals etc). However, the amount of rangeland required to support the livestock which might supply manure to compensate for losses from arable lands is less in the higher potential zones.[14]

While there has undoubtedly been a range of high-quality scientific research on soil management questions in Africa over the last century, resulting in some important conclusions, in order to look behind the neat statistics and apparently concrete results, we must interrogate the methodological assumptions used in mainstream analyses of soil change in Africa by exploring the styles of investigation conventionally used. This is the subject of the next section.

STYLES OF INVESTIGATION AND SOURCES OF EVIDENCE

The methods used by scientists to understand soils have naturally changed along with the foci of research described above. Several approaches have been important in framing the way we understand Africa's soils. Below, three broad categories of methods are discussed: surveys and classification; controlled experimentation on plots; and nutrient budget analyses. As lenses through which mainstream soil science has looked at the issue, such approaches have had enormous influence over the way problems have been defined, and potential solutions elaborated. The selective use of such findings has been key to the sort of policy proclamations introduced earlier.

Surveys, classifications and plans have been enormously influential in structuring the way agricultural experts and planners have viewed soils in Africa. Continental or national soil maps, for instance, divide areas into different categories according to the key classifications. At a more local scale, different parts of a country or region may be classified according to the suitability for different land uses. The associated discipline of land-use planning has often made good use of soil surveys to design plans and reshape agricultural landscapes along lines deemed to be technically most appropriate. But, in attempting to create a stable, universal ordering, conventional soil classifications and land-use plans are necessarily reliant on certain stable features of soils and landscapes, and take scant notice of local variations or dynamics. The result is that fine-tuned local classifications used by farmers are ignored, and an aggregate pattern is imposed. This has had major consequences in each of the case study countries, with land-use planning (based in large part on soil mapping) being a significant input into centralization and land husbandry policies in Zimbabwe from the 1930s, the villagization schemes in Ethiopia in the 1980s, and the planning of the cash crop zones in Mali.

The problem of aggregation through standardized classifications is particularly apparent when we examine the results of large-scale assessments of soil degradation in Africa. There have been a number of attempts – at continental, national and regional levels – to assess such issues as erosion hazard, erosion incidence, soil degradation or desertification. The maps produced from such surveys have enormous influence, and become powerful tools in policy advocacy, framing the way interventions under such initiatives as the Convention to Combat Desertification are thought about. For example, as part of the follow up to the UN Conference on Desertification held in Nairobi in 1977, UNEP (the United Nations Environment Programme) commissioned a review of desertification based on a questionnaire survey sent to 91 countries (Swift, 1996). The study concluded that 'desertification threatens 35 per cent of the Earth's surface and 20 per cent of its population' (UNEP, 1984, p17). Similar statistics emerged from the Global Assessment of Soil Degradation (GLASOD) study which concluded that some 26 per cent of the dryland areas of Africa were suffering from some degree of soil degradation, and, across the continent, nearly 500 million hectares of land were degraded (Oldeman et al, 1990; Oldeman, 1994). Such statistics have a major influence on the imaginations of politicians and publics alike, and despite the nature of the data from which they are derived, have huge sway in policy debates (Swift, 1996).

Experimental plots have been the most important source of specific biophysical information on African soils. Especially when parameters have been monitored over considerable periods, these have revealed important information about soil-fertility change under different management regimes. Similarly, controlled experiments to understand patterns of soil loss, nutrient limitation and yield response under different conditions have been important in designing soil conservation and fertility input regimes.

However, such data have clear limitations. First, the particular conditons of research stations may not reflect the wider farm setting; often research stations have better water and soil conditions and the management regimes imposed may not reflect farmers' own realities. Unfortunately most experiments have been under research station conditions, with relatively few being undertaken by farmers themselves or even in field conditions.[15] Second, the time depth of most experimental observations is limited. There are some notable exceptions, of course, but of the 21 long-term experiments reviewed by Swift et al (1994), only three spanned a period of 20 or more years, making it difficult to assess longer-term dynamics given the variability of climate and soil change in African settings. Third, plot-based data cannot be extrapolated to wider areas. What happens on a plot may not happen on a larger area due to different dynamics occurring at wider spatial scales. This is particularly so for soil erosion data, because soil lost from one part of the landscape may not be permanently lost, but simply redistributed. Thus, extrapolating a total soil loss figure from individual plot level is erroneous and misleading (Stocking, 1987). Finally, controlled experiments, by attempting to eliminate variability and control variables for statistical analysis of treatment comparisons, may miss out on key insights. By choosing standardized, levelled plots, by making management inputs uniform, and by eliminating data which is seen to be not part of overall trends, critical aspects of real-life variability and complexity may be hidden from view by conventional experimental design and analysis techniques. While there is now more discussion of alternative statistical analysis which takes variability seriously (Riley and Alexander, 1997) and methods for experimental design which capture the dynamics of micro-variation (Brouwer et al, 1993), this remains peripheral to mainstream scientific practice.

Nutrient-balance assessments have increasingly become another important methodological tool for looking at soil-fertility issues, as illustrated in Table 1.1. Some of the earliest attempts in the African setting focused on continental or regional scales (Stoorvogel and Smaling, 1990). The continental assessment, in particular, had a major impact on thinking about soil-fertility management, and, as we saw earlier, the figures are widely quoted by scientists and policy-makers alike. Drawing on this to make the case for significant new public investments in soil-fertility management, Sanchez et al (1997, p1) state:

> *Soil-fertility depletion in smallholder farms is the fundamental biophysical limiting factor responsible for the declining per-capita food production of sub-Saharan Africa. The magnitude of nutrient mining is huge. We estimate the net per-hectare loss during the last 30 years to be 700 kg N [nitrogen], 100 kg P [phosphorous], and 450 kg K [potassium] in about 100 million hectares of cultivated land.*

More recent efforts have concentrated on smaller scales, such as the farm, plot or niche (see Table 1.1). Essentially the methodologies used are the same: all inputs (from inorganic fertilizers, organic manures/composts, crop residues, atmospheric deposition, soil run-on, nitrogen fixation etc) and all outputs (from harvesting/grazing, crop residue removal, leaching, gaseous loss and soil erosion) are measured or estimated. By calculating the amount of nutrients (usually nitrogen (N) and phosphorous (P), and sometimes potassium (K)) in each of the materials, a nutrient balance can be calculated for the area being investigated. While such balance studies have considerable heuristic value as a way of thinking about the efficient management and conservation of nutrients in an integrated way (Defoer et al, 1998b), the data derived has some inevitable problems because of the crude nature of the analysis, with particular dangers when applied to broader policy analysis (Scoones and Toulmin, 1998). A number of problems have been commented upon.

- First is the issue of estimation error. Nutrient budgets may be derived from actual measurement, transfer functions and literature estimates. When combined, errors may accumulate resulting in estimations which must be subject to careful sensitivity analyses (Smaling and Oenema, 1997).
- Accounting models of this sort necessarily make certain assumptions about underlying processes. A black-box approach to internal soil dynamics is taken, with the concentration on input and output flows. This ignores the possibility of key aspects of soil-fertility being influenced, not by nutrient balances per se, but by other aspects of the soil–plant interaction (Noordwijk, 1999).
- As with other approaches, attention to scaling issues is important. Patterns of nutrient balance may be quite different at different scales, and differentiation between niche, plot, farm and wider scales needs to be made before generalizations based on unwarranted extrapolation are made. At larger and larger scales nutrient balance levels would, from first principles, be expected to tend towards zero, as nutrients get redistributed. At a global level, for instance, nutrient losses and gains are expected to be effectively in balance, while at smaller scales greater variability between sites would be expected, with some exporting and others importing nutrients. Although the smaller-scale studies certainly show high levels of variation in balance estimates between different niches, plots and farms (see Table 1.1), the larger-level studies (at regional and continental scales) do not show the expected pattern. This seems to be partly due to basic scale errors, because the data used are aggregated up from nutrient exporting sites (ie arable fields), and so do not account for nutrient deposition elsewhere.
- Nutrient budgets give a snapshot assessment of the balance of current flows of nutrients. They do not give any indication of how this relates to the overall stocks of nutrients available, nor the broader trends in balance levels for a particular case. Thus while a negative balance is clearly not wonderful news, it may not be as calamitous as is sometimes suggested. In some cases nutrient depletion is occurring in settings where considerable stocks exist, and no immediate concern for productivity is apparent. In other cases, nutrient depletion may be the most sensible option in the short to medium term, if, over the longer term, under changed economic or social conditions, investment in soil improvement then takes place (Scoones and Toulmin, 1999).

As the case study chapters show, nutrient balance studies can provide useful insights if firmly located in field-level realities, with the appropriate caveats added and other contextual information provided. As a field-level management tool to encourage discussion about different options, the approach has proved most valuable (Chapter 5). However, as with surveys and experiments, if used in an unreflective manner, particularly when extrapolated to broader scales, the nutrient balance approach can be highly misleading. For this reason it is important to interrogate a bit further the underlying assumptions of current research practice in order to develop new ways of looking at the issues.

SCIENCE AND HOW THE POLICY DEBATE IS FRAMED

As we have seen, soil surveys and classifications, experimental plot measurements and nutrient balance studies have important embedded assumptions about soils and their dynamics and, in adopting particular sets of methods, ensure that the world is seen in a particular way. Such perspectives are not necessarily 'wrong' or 'inaccurate' in any objective sense, but, as discussed above, they must be seen as necessarily partial and limited. With problems framed in a particular way, particular solutions necessarily emerge. The panoply of soil management interventions – from soil amendment recommendations, to soil conservation measures, to the integrated soil-fertility management packages discussed earlier – emanates from a set of scientific understandings, derived from a particular history of enquiry.

Over the past century an identifiable diagnosis of problems and solutions has therefore emerged. This 'narrative' has a number of key elements, each significant in framing the policy debate. First, there is near-universal consensus that soil degradation is a significant and growing problem in Africa, requiring urgent action lest yields decline and potential starvation and social unrest result. Second, a set of technical solutions is advocated to rectify the situation. The emphases vary, with some advocating solutions more focused on inorganic fertilizers, while others argue for a more organic approach. The emerging middle ground – typified by the integrated soil-fertility management approach – is perhaps the most common today. These technical solutions combine to make up the third element of the narrative, which sees them combined as part of an idealized, settled, mixed farming system, replacing 'backward' shifting cultivation or transhumant pastoral systems. In the mixed farming model, crops and livestock are integrated, soil nutrients are recycled and modern technologies are applied to improve effiencies under a system of exclusive land tenure (McIntire et al, 1992; Winrock, 1992).

As the earlier discussion has shown, elements of this are easily identifiable in contemporary policy statements on the African environmental situation (see also Chapters 2 to 4 for country-specific commentaries). But such an argument has not emerged recently. Indeed a narrative derived from the diagnosis of environmental crisis, leading to the need for the development of an efficient, modern, mixed farming model based on a series of fairly standard technical recommendations can be traced back at least to the 1930s (Sumberg, 1998; Wolmer and Scoones, 2000; Scoones and Wolmer, forthcoming). Alarm about the prospects of large-scale environmental degradation was in particular prompted by the widespread droughts of the 1920s, and the experience of the US dust bowl in the 1930s (Anderson, 1984; Beinart, 1989). The proposed solution centred on a combination of mechanical soil conservation and fertility management, particularly through organic matter management, rotations and leys, all combined as part of an integrated mixed farming model based on the long-established European system. By the 1940s, across colonial Africa, research and policy were increasingly focused on this range of technical interventions. For example, in Nigeria demonstration farms to show the benefits of mixed farms were established

(Tempany et al, 1944), while in Zimbabwe major land-use reorganization and soil erosion efforts got underway (Chapter 4).

This basic narrative of the problem and the associated implied solutions has become deeply embedded in the assumptions of scientists, policy-makers and others, and is continuously reinforced by institutional settings. It is therefore not surprising that the basic features persist today in largely similar form, and continue to have a major influence on policy thinking. The concern of our research, however, is not to dispute each of the elements of the argument. Many are sound when applied to particular settings. The important point is to recognize that such views are necessarily limited and partial. The key question is: given other assumptions, alternative methods and different types of analysis, would the world look different, and – most importantly for practical development and policy – would alternative policies and strategies be suggested? In the next section these questions are pursued in some detail. First, however, it is necessary to dissect the key tenets of mainstream analyses. A number of themes are evident.

First is the disciplinary focus of most mainstream research, derived almost without exception from natural scientific concerns. At different times, different natural science disciplines have dominated – pedology, soil physics and chemistry, soil biology and ecology, experimental agronomy and so on. But ultimately a technical perspective has prevailed. Soils are understood in terms of nitrogen or phosphorous content, cation exchange capacities, water holding capacity, microbial biology and so on, but social and economic perspectives have been very limited, and if present certainly marginal.[16] As Swift (1998, p59) observed at the 16th World Congress of Soil Science:

> *Soil science has been brilliantly informed by reductionist physics and chemistry, poorly informed by ecology and geography, and largely uninformed by the social sciences.*

While there has been some interesting research on local soil classifications and the links with scientific classifications (Talawar, 1996; Kanté and Defoer, 1994), this has had only limited impact. Other social science work has failed to engage with technical and policy issues almost completely, concentrating instead on the social, cultural and symbolic interpretations of soils and their fertility (see Jacobsen-Widding and van Beek, 1990). Only in a few rare cases have the social and the natural science issues been brought together to attempt a more integrative analysis.[17]

Second, and deriving from the disciplinary focus of most research, is the technology-centred approach to intervention. Huge numbers of technologies and management recommendations have been derived from scientific research over the last century, across a wide range of areas. But most of this work has focused on achieving an optimal agronomic solution. The assessment measures have been technical – yield, soil loss, nutrient levels and so on – and not necessarily rooted in a social, political or economic understanding of agriculture and environmental management among a highly differentiated farming population. Where economic analyses have been made – for example in relation to fertilizer application rates – this has certainly been an important advance from simply looking at technical parameters.[18] But relative marginal returns may be only one decision criterion for a farmer, and choices may be conditioned by a range of other social and institutional factors. Much research now shows how the socio-economic conditions for successful soil-fertility management may be just as important as technical factors (see Scoones and Toulmin, 1995). As the case study chapters amply demonstrate, constraints on access to land, labour and capital may be influenced by a range of formal and informal institutions including input and output markets, resource tenure, gender relations, labour provisioning, and so on. Yet such insights rarely become integrated into the technical research which dominates the soil-fertility research agenda.

The technical focus of most research on soils in Africa, in turn, influences the definitions of land degradation used by most analysts. Land degradation is an emotive and ultimately normative concept, carrying with it, as we have seen, significant policy ramifications. Understanding what land degradation is (and is not) is therefore a critical area. Most assessments of land degradation, however, take a purely technical line: if soil is being eroded or nutrients are being lost, this constitutes land degradation. The indicators of degradation are therefore the ones being measured by mainstream technical science – soil chemical properties, erosion loss, nutrient balances etc. A sense of objectivity and rigour is created, but do such studies necessarily measure 'degradation' in a broader sense, or just some processes of biophysical change? A more robust definition of degradation accepts that it is necessarily a normative concept and must be related to the social, economic and other values (both future and present) associated with the soil resource.[19]

The key question, then, is: do the observed changes in soil chemical properties, erosion levels or nutrient balances matter? This question refocuses our attention on the use of soils for people's livelihoods (as well as for broader societal benefits, such as carbon sinks or the hydrological cycle). There are therefore occasions when negative biophysical changes (usually referred to as 'degrading') are not problematic and so should not be categorized as land degradation, if the definition proposed here is embraced. For example, the impacts of soil depletion on people's livelihoods may be limited when there are low rates of extraction or extensive reserves; when substitutes for natural soil capital exist; or when alternative livelihood sources exist which reduce the dependence on the soil resource (Scoones and Toulmin, 1999).

Finally, the methodological stance used conventionally in mainstream scientific investigations has important ramifications. Some of these problems have already been mentioned. Design and analysis of experiments is inevitably influenced by taking an essentially ahistorical approach; paying limited attention to multi-scaled spatial diversity and complex temporal dynamics; adopting usually a rather linear interpretation of environmental change; and using normal distributions and means in most statistical analysis. The net result is that a linear, undifferentiated and technical perspective is projected which hides from view much of the diversity and complexity of soils as they are managed by real farmers.

While elements of the mainstream technical perspective are of undoubted use, an alternative, complementary perspective on soil management in Africa is opened up by adopting a somewhat different conceptual and methodological stance. The key elements of the approach are outlined in the following sections, drawing on a summary of the findings from the case study research.

CASE STUDY SITES: SOME CONTRASTS AND COMPARISONS

The teams involved in the case studies presented in Chapters 2–4 have focused on a range of different sites in Ethiopia, Mali and Zimbabwe. The aim has been to explore, through locally-based research with farmers, the complex diversity and dynamics of soil-fertility management in different small-scale farming settings in Africa, attempting to shed light on broader policy debates by linking understandings of local-level processes with broader macro-policy change (see Chapter 6). An interdisciplinary team-based approach to the research was adopted, involving both natural and social scientists working in collaboration with farmers.[20] An important starting point was discussions with farmers on their own understandings of soils and soil-fertility change. This involved village and farm-level mapping according to local soil classifications. This was complemented by discussions of field, plot and broader landscape histories. A key element involved resource flow mapping, exploring how different materials move in and out of different parts of a farm and how they influence the changing status of soils according to local criteria. Such farm-level discussions helped frame the questions for subsequent investigations by both natural and social scientists, as well as helping to set an agenda for participatory action research on particular problem areas identified (see Chapter 5).

Natural science investigations focused on the flows of nutrients (particularly N, P and, in some cases, K) in and out of the whole farm system and its different sub-components across a range of case examples stratified according to local classifications of wealth or soil management capability. The nutrient budget analyses (see below for a further discussion) which emerged were able to fill in details within the farmers' own resource flow maps with information on soil nutrient status, and the nutrient contents of different materials. The questions pursued in the parallel social science investigations concerned the social, cultural, economic and political factors which influenced the various flows and stocks of nutrients at a farm level. For example, the examination of the institutions governing labour relations within and between households highlights the socio-economic processes influencing particular flows of fertility resources. Similarly, economic analysis of prices and markets offers insights into the relative incentives for different options. Examinations of resource tenure, in turn, highlight how tenure regimes and perceptions of security influence the management of nutrient stocks and soil-fertility.

A particular emphasis in each of the case studies was to explore changes over time to set an understanding of the contemporary situation in a historical context. Unravelling the complex interactions between social, economic and political change and patterns of soil-fertility in farmers' fields is no easy task. However, a combination of oral histories, archival records and time-series data was pieced together for each of the sites to give at least a schematic picture of trends and processes over time. This work highlighted the importance of policy contexts for soil-fertility management, as invariably the historical enquiries emphasized the importance of the combination of the impact of external events and local processes for changes in land use and management.

Thus over a period of several years a detailed picture has been built up of the interconnections between biophysical processes of nutrient accumulation and depletion, and a range of socio-economic processes operating at the local, national and sometimes international levels. The result, as will be evident from a reading of the case studies presented in Chapters 2 to 4, is a highly complex story, one that is a far cry from the generalized picture presented in much of the policy debate over recent years.

The case study sites in Ethiopia, Mali and Zimbabwe represent a wide range of settings in the savanna farming zones of east, west and southern Africa. While the study clearly cannot claim to be representative of all such farming situations in Africa, some important contrasts are highlighted both in terms of biophysical conditions and broader socio-economic and policy contexts. By choosing sites differentiated by agro-ecology and case study farms according to socio-economic criteria, a comparative analysis which contrasted relatively high and low potential areas, and relatively richer and poorer farmers, was made possible. Such comparisons can be made at a number of levels: between countries, village study sites and particular farms or plots.

Contrasts between countries

The case studies examined in this book are all located in the small-scale peasant farming sectors where poverty levels run high. Real GDP per capita (1995 figures) for Ethiopia, Mali and Zimbabwe was US$455, $565 and $2135 respectively (although considerably less in the communal areas), whilst the countries were ranked 169th, 171st and 130th respectively out of 174 countries according to the composite Human Development Index (UNDP, 1998). In all three countries, agriculture is the major contributor to the national economy, both through providing a subsistence base for much of the population, and cash incomes and export earnings through more commercial farming. The structure of the agricultural economy in each of the countries has been highly influenced by past policies. In the case of Zimbabwe, for instance, a dual economy is evident, with large-scale commercial farming on previously exclusively white-owned land existing alongside small-scale agriculture in the communal areas. In Mali a major difference exists between those areas within the large cash-cropping zones established during the colonial era to encourage the production of cotton or rice, and those outside where more extensive, marginal dryland farming and livestock keeping is evident. In

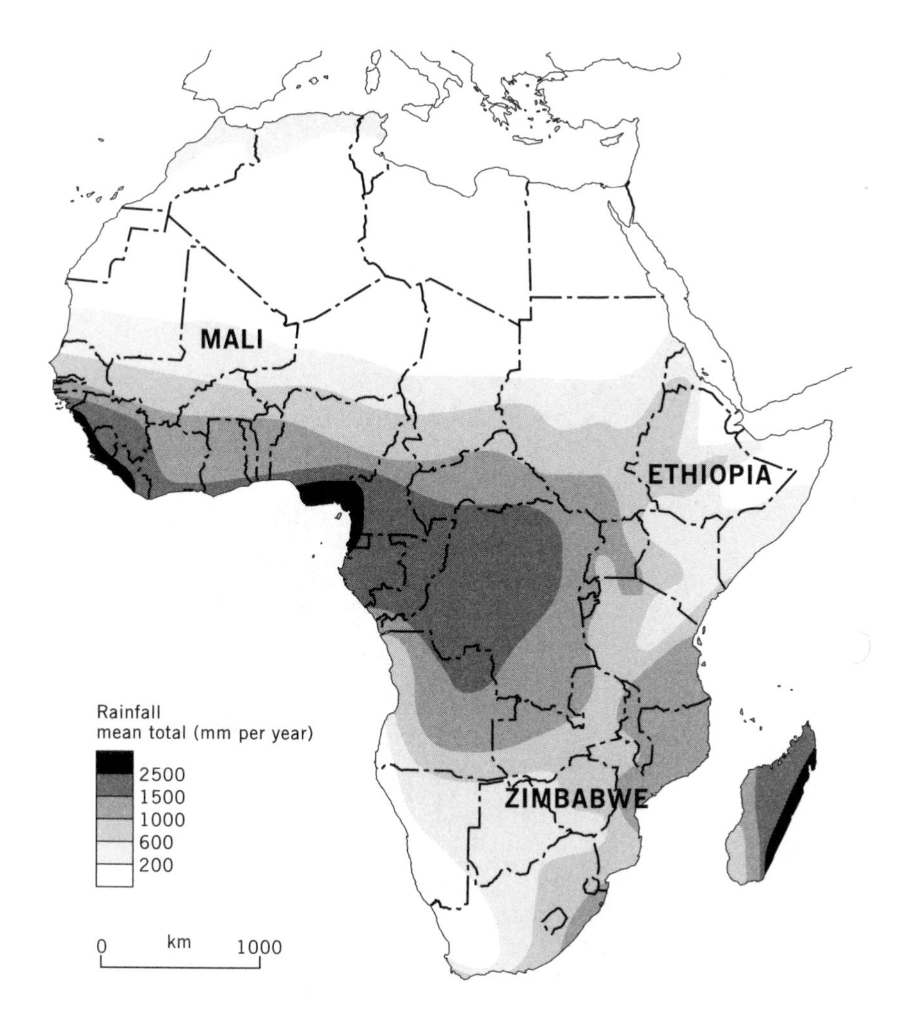

Figure 1.1 *Rainfall distribution in Africa (mean total rainfall per year)*

Ethiopia government policies have again had a major impact on agriculture, with many areas outside the major grain producing zones and the previously state-owned farms receiving limited attention.

Figure 1.1 shows the location of Ethiopia, Mali and Zimbabwe in relation to the distribution of rainfall across Africa. All fall within the Sahelian and savanna zones, where natural vegetation consists of a mix of tree and grassland. Annual rainfall ranges from around 350mm at the driest end to around 1250mm at the wetter end of the scale, with high levels of interannual variability in all sites.

Some major contrasts are apparent in the soil types and associated geology (see Figure 1.2), as the three countries encompass the major soil groups represented in Africa. In Mali, ancient weathered sands (oxisols, lithosols and acrisols away from the drier parts of the Sahel and Sahara) dominate. These are severely

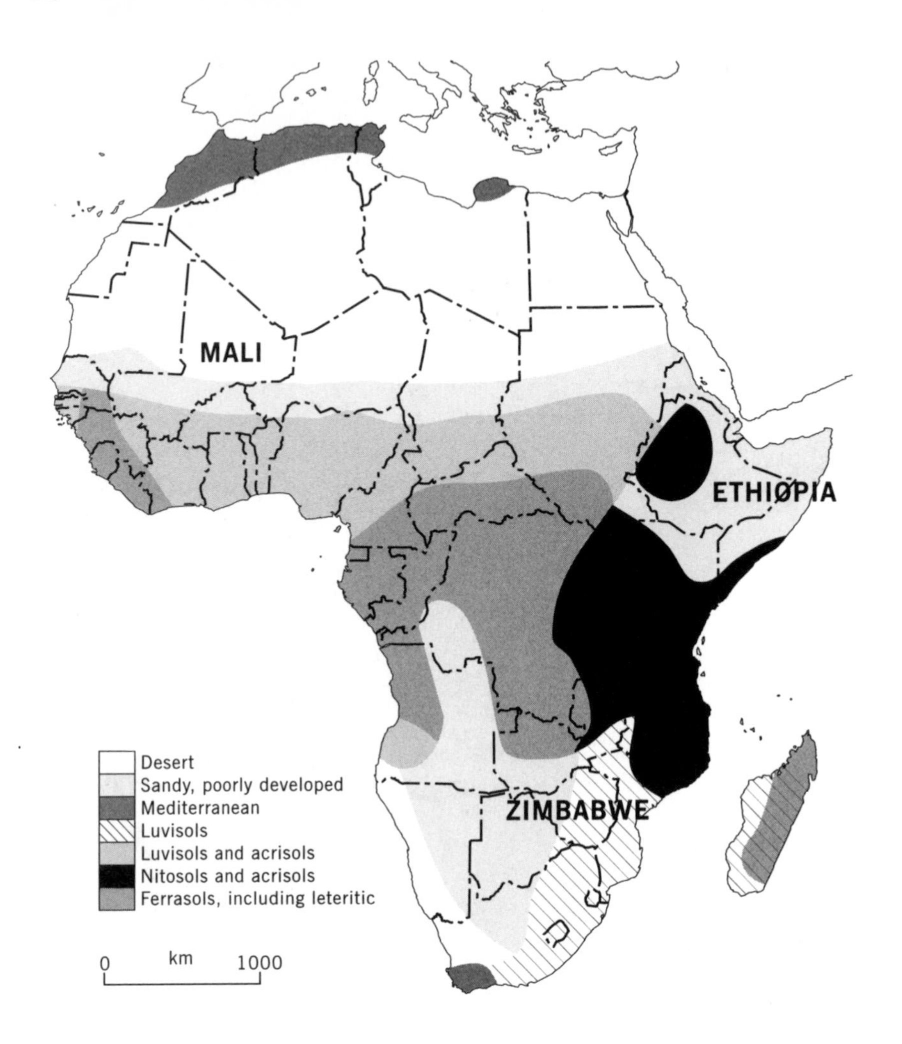

Figure 1.2 *Soil groups in Africa*

deficient in mineral nutrients, have low clay and organic matter levels, very often have a poor water-holding capacity, may be subject to acidification with low cation-exchange capacities, and, in more sloping areas, are subject to erosion with much resultant variation across toposequences (see Chapter 3). However, in contrast to other areas, soils in the savanna zone of west Africa benefit from extensive deposition of nutrients from dust deposited during the Harmattan (de Ridder and van Keulen, 1990; van Duivenbooden, 1992; Pieri, 1989). Sandy soils (luvisols) derived from granite also characterize the Zimbabwe sites (Thompson and Purves, 1981). These are generally deficient in N, and sometimes P and other micronutrients (Grant, 1981). They also have poor water-holding capacity and can be subject to significant erosion (Elwell, 1985; Whitlow, 1988). The Ethiopian sites, by contrast, are nitosols derived from relatively recent volcanic material (Weigel, 1986). These are comparatively nutrient-rich and have high clay contents. P fixation is a problem, particularly in the highland soils where it may become limiting to plant growth (Belay, 1992). The high proportion of sloping land, particularly in the highland areas, means that these sites are prone to erosion (Hurni, 1994).

In recent years a range of national government policies have had major impacts on the agricultural sector (see Chapter 6). Since the late 1980s, in all countries, structural adjustment policies have resulted in various forms of liberalization with major effects on input and output prices and marketing, the provision of rural services, agricultural extension, opportunities for off-farm employments, and urban–rural remittance flows.[21] Land reform, land-use planning and resettlement policies have also been important, particularly in Ethiopia and Zimbabwe, where over time various attempts have been made at villagization, land redistribution and resettlement. Land management and agriculture have also been affected by decentralization policies across all countries, although the character of such policies and their influences differ (see Chapters 2 and 4 for details).

Contrasts between study sites

In the design of the research, study sites were chosen to capture a range of important national or regional contrasts. Thus in each country a series of sites (two in Ethiopia and Zimbabwe and four in Mali) were chosen along a transect running from relatively high to low resource endowment. In the case of Ethiopia, the research focused on one region, North Omo, in the Wolayta enset-root crop-based system, with both a lowland and highland site. In Mali, in addition to two dryland agropastoral sites (Dilaba and Siguiné), two cases were added to explore the irrigated rice (Tissana) and cotton zones (M'Péresso). In Zimbabwe, the sites are found in two communal areas, one in the higher potential part of the country (Mangwende) and the other in the drier zone (Chivi). Table 1.2 provides a summary of some of the key contrasts between sites, including both agro-ecological and socio-economic characteristics. Aspects of these contrasts are discussed in the following sections.

Agro-ecological contrasts

The options for soil-fertility management in each of the sites is critically dependent on the interaction between plant-available nutrients and soil moisture. The inherent fertility of the soil, combined with the history of soil management, affects available nutrients, while patterns of rainfall, soil texture and structure, and the management of water within fields through water conservation and harvesting techniques, affect levels of soil moisture.

In different sites at different times, either nutrients or water are limiting to plant production. In savanna ecology, a useful general distinction between savanna types is made. These include 'eutrophic' areas with clay rich soils and low infiltration rates where, especially in the drier areas, soil moisture is limiting; and 'dystrophic' areas with poorer sandy soils and high infiltration rates, where soil nutrients are limiting, especially in the wetter areas (Frost et al, 1986; Menaut et al, 1985; Scholes, 1990). For example, the rich volcanic soils

Table 1.2 *Key contrasts between study sites*

	Ethiopia		*Zimbabwe*		*Mali*			
	Highland	*Lowland*	*M'wende*	*Chivi*	*Dilaba*	*Siguiné*	*Tissana*	*M'Péresso*
Rainfall (mm)	1272	924	850	550	450	450	650	800
Major soil type	Nitosol (clay)	Nitosol (clay loam)	Granitic sands	Granitic sands	Lithosols, acrisols (sands, gravels)	Acrisols (sands, gravels)	Lithosols, acrisols (sands, gravels)	Lithosol, acrisol, gley soil (sand/ loamy sand
Major agricultural focus	Enset and root crops	Maize, cotton	Maize, cotton, sunflower	Maize, small grains, groundut	Dryland cereals	Dryland cereals	Irrigated rice	Cotton, cereals
Population density (people/km^2)	375	110	150	44	50	15	29	18
Ethnic composition	Wolayta	Wolayta	Shona	Shona	Bambara	Bambara	Diverse, drawn from elsewhere in west Africa	Minianka

of the Ethiopia study site, particularly in the highlands, contrast dramatically with dystrophic systems of the poor, weathered soils of the Mali sites and the granitic sands which dominate the Zimbabwe sites. Such diverse characterizations make any generalizations about 'African soils' highly problematic. Figure 1.3 attempts to locate the different study sites across the two axes of plant available moisture and nutrients.

Thus both the Ethiopian sites lie in areas of relatively fertile volcanic soil, and a simple rainfall gradient distinguishes the highland and lowland sites. In Zimbabwe, both sites are found on poor granite sands, but these have much lower inherent fertility in the higher rainfall site of Mangwende, due to leaching and intensive use in the past. The poorest soils of all, in terms of nutrient content, are found in the Sahelian sites in Mali (Siguiné and Dilaba) which are again found along a rainfall gradient. The other Mali sites have poor soils, although better water availability through higher rainfall in the case of M'Péresso and irrigation water in Tissana (see Chapters 2–4 for more details).

While such simple contrasts hide a great deal of variation within sites (see below), they do highlight how agro-ecological dynamics, and associated strategies for soil management, differ. Thus for those sites, such as the highland site in southern Ethiopia, found towards the top left of Figure 1.3, higher nutrient stocks and a relatively slow release of nutrients are evident, although productivity may be constrained by immobilization, erosion and leaching. Under these conditions, strategies for increasing soil-fertility in the long term through sustained application of inputs are possible, as residual benefits can be captured, by the building up of soil-fertility in areas such as the enset and taro gardens in highland Wolayta (see Chapter 2).

By contrast, for those sites found towards the bottom right of Figure 1.3, a different dynamic is expected. Here, limitations on productivity due to the lack of both water and nutrients may apply as a result of low organic matter levels, inherently low nutrient levels in the soil and limited water-holding capacity. Variability in rainfall is a significant ecosystem driver, with pulsed release of nutrients, intermittent erosion events, and shifts between water and nutrient limitation across years and between seasons. In such settings a much more opportunistic soil management strategy is required, with attention paid to the boosting of nutrient use efficiency and the timing of soil-fertility management activities.

For those sites with better water availability but poor soils, the key challenge is to increase available soil nutrients through increasing inputs. However, the dangers of erosion, leaching, acidification, rapid decomposition and mineralization may offset such efforts. In such sandy soils, frequent additions of high quality organic matter and mineral nitrogen are required (Buresh and Smithson, 1997), but also attention to other mineral components (eg K and P) is necessary where the residual benefits of application are relatively low.

The patterns of interaction between soil moisture and nutrients also vary hugely over time. Between years, for instance, changes in rainfall levels may result in shifts between water and nutrient limitation within in a particular site, and make different soil niches more or less productive. For example, with the sustained decline in rainfall since the 1960s across the Malian sites, the previously highly-valued, heavier, relatively nutrient-rich soils in the dryland areas have become increasingly less productive because of lack of soil moisture, while the sandy soils with good infiltration properties and the valley bottom

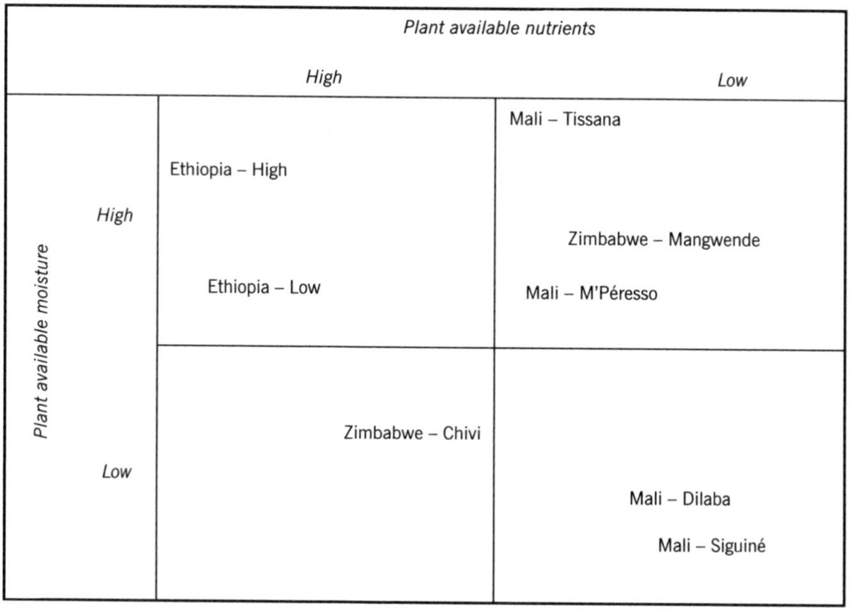

	Plant available nutrients	
Plant available moisture	*High*	*Low*
High	Ethiopia – High Ethiopia – Low	Mali – Tissana Zimbabwe – Mangwende Mali – M'Péresso
Low	Zimbabwe – Chivi	Mali – Dilaba Mali – Siguiné

Figure 1.3 *Contrasts in savanna ecology across sites*

areas (*bas fonds*) are now regarded as more valuable (see Chapter 4). Even within fields such variations may have major effects on productivity, with micro-variations making either soil moisture or soil nutrients limiting in a highly variegated manner (Brouwer and Bouma, 1997; Brouwer and Powell, 1998).

Within a season, there may be high variations in the availability of particular nutrients, especially nitrogen, due to the complex interaction of soil chemical, physical and biological processes (Scholes et al, 1994). Especially at the onset of the rainy season, soil wetting results in increased mineralization and nutrient release (Semb and Robinson, 1969; Frost, 1996). For example, in the dryland sites of Mali and Zimbabwe, this results in early flushes of natural vegetation and the opportunity for dry planted crops to capture nitrogen in their early growth phases. However, such effects are often counteracted by leaching, denitrification and immobilization so that the nutrients actually available for plant growth are hugely variable (Buresh and Smithson, 1997).

Thus depending on the location and the time, different niches within the wider study areas may be located in different quadrants of Figure 1.3. This, of course, has implications for soil-fertility management, as a standard, blanket approach across space and time is clearly not appropriate. As the case studies clearly demonstrate, there is an enormous amount of spatial and temporal diversity in soil properties, and so in management strategies.

Socio-economic contrasts

A range of socio-economic characteristics also influences options for soil-fertility management across the study sites. As shown in Table 1.2, population density varies from 375 people per square kilometre in highland Ethiopia to only 15 in Siguiné in the dryland Sahelian zone of Mali. Widely differing land to labour ratios have major implications for patterns of agricultural management and processes of intensification. In highland Wolayta, land is at a premium with plot sizes averaging only 0.6ha, while in Siguiné land is relatively abundant, with household holdings averaging 44ha. Thus incentives to invest in labour-intensive soil management activities vary dramatically across sites, with highly labour-intensive gardening typifying the highland Wolatya site (see Chapter 2), while more extensive, low-input bush–fallow systems are more typical in the Malian Sahel sites. The degree to which agriculture or pastoralism are central to people's livelihoods also varies with population densities. In the highland Ethiopia case, for example, the small land holdings mean that survival from land-based production alone is insufficient, and other off-farm activities must be added to a wider portfolio of activities (Carswell et al, 2000). By contrast, in more extensive systems land areas may be sufficient, although high levels of risk may require alternative income-earning options.

Such patterns are, of course, in flux. Population increases are evident in all sites, with national and regional averages of around 2–3 per cent.[22] In all sites this has brought about a shift in livelihood portfolios – towards trading, craft work, or migration to towns or commercial farms. With such changes, the incentives to invest in soil management on the home farm will also alter. For example, in Zimbabwe circular migration has long been a feature of the rural economy, meaning that, although population pressures are relatively high given the agro-ecological conditions, the availability of alternative souces of income through remittances has offset the incentives to invest in agriculture and soil improvement for many. As Chapter 4 indicates, this may now be changing as shifts in the broader economy following structural adjustment have reduced real wages and led to a contraction in employment opportunites. Now many male communal area residents, who previously would have worked away, are investing in agriculture and soils at home in the communal areas. In the dryland site of Dilaba in Mali, limits to the extensive bush–fallow system are being felt as the village fields have extended to the edge of their territorial boundaries. Here, too, changes in farming and soil-fertility management strategies are evident, with greater investment in home fields and *bas fonds*. The key constraint here is the availability of manure, as grazing land for livestock is increasingly scarce. In Siguiné, by contrast, fallowing remains an option, at least for the time being (see Chapter 3).

The relative availability of land and labour are, of course, not the only factors influencing patterns of land intensification and incentives for soil-fertility management. The relative price of inputs and outputs is another important consideration. This is affected by a range of factors including, among other things: marketing and pricing policy, the location and type of input and output suppliers, traders and markets, and the quality and effectiveness of transport infrastructure (see Chapter 6). As discussed at length in the case study chapters, such conditions vary considerably across sites. At one extreme are the sites located within the cotton and rice areas of Mali, where parastatal-supported output marketing and input supply has encouraged the widespread and generally profitable use of fertilizer on cotton and rice. Long-term investment dating back to the 1920s has also ensured that such areas are well provided with infrastructure and other support (Chapter 3). At the other extreme lies the more remote lowland site in Ethiopia which, until recently, had no year-round road access and, with the exception of a period during the 1970s when a large integrated rural development project operated in the area, the site has had poor input supply and adverse terms of trade. This has made inputs expensive relative to the prices offered for crops, with the result that investment in fertilizer, for example, has been highly constrained for most farmers (Chapter 2). Broader service support also influences options for soil-fertility management. For example, access to credit and information from extension services may be critical factors in the adoption of particular soil-fertility management options. This is particularly important for the adoption of inorganic fertilizer, given its often high cost and the skills required for effective application. Studies in Ethiopia, for example, have shown how fertilizer use is highly correlated with both access to household assets and access to services, notably credit, agricultural extension and school education (Croppenstedt et al, 1998; see also Chapter 2). Key knowledge and skills, combined with the willingness and ability to experiment with new soil-fertility management options, are seen to be important in the Mali case study (Chapter 3). In some parts of the country, support for farmer groups and processes of monitoring and experimentation have reinforced farmers' own abilities to manage soils (see also Chapter 5).

Land tenure security is often mentioned as a key factor influencing the likelihood of technology adoption and investment in environmental management. Across the sites, however, relatively secure de facto land and resource tenure is evident, and the empirical studies suggest that existing patterns of resource tenure are not a significant constraint to investment in soil-fertility. However, this has not always been the case. In Ethiopia, past policies of land reform and villagization, for example, have introduced a great sense of insecurity with the consequence that farmers desisted from investing in longer-term assets, such as soils and trees, for fear of forced expropriation or resettlement. Vestiges of this lack of trust in the state are evident today, but, by and large, the field evidence suggests that farmers have returned to investing in gardens, trees and other long-term productive resources under a variety of complex tenure settings, ranging from de facto private ownership to various contracting and sharecropping arrangements (Chapter 2).

Broader cultural factors may also have an influence on attitudes to soil management and the strategies pursued. As the case study chapters show, local understandings of soils are deeply embedded in socio-cultural institutions. Practical knowledge about soils and their management is related to people's understanding of the relationships between resources and their fertility. In Ethiopia, for example, the fertility of soils is seen to parallel interpretions of human health and fertility (Data and Scoones, forthcoming). Similarly, in Zimbabwe, the status of soils in a particular farm is seen to relate to a wider spiritual realm, with good results arising only if appropriate actions in relation to spirit ancestors are taken (Chapter 4). Farmers' practical knowledge of soils and their management is thus deeply entwined with social relations and broader cultural understandings of the relationship between human, spiritual and natural worlds.

Contrasts within sites and between farms and plots

Within sites, there are also important spatial variations. Not all sites have a uniform geological origin. For example, in Zimbabwe, doleritic intrusions in a wider soil landscape of granitic origin mean that patches of heavy clay soil with eutrophic properties are found within the more widespread dystrophic sands. Variations also typically occur across slopes, with catenas showing variations of soil type from hill top to valley bottom. Site topography, therefore, has important implications with fields found higher up the slope typically being drier and with poorer soils, while lower slope, riverine and valley bottom areas may be particularly significant 'key resources' in the agricultural system (Scoones, 1991). For example, river banks and valley bottom *dambos* in Zimbabwe provide important sites for gardens. In these locations, available soil nutrients and moisture are significiantly higher than the surrounding areas, opening up options for sustained investment in soil-fertility improvement which are largely impossible elsewhere (Scoones and Cousins, 1994). But not all variation in soil properties is the result of underlying geology or the consequences of topography; historical legacies of past practices also add to the variable spatial patterning of different soil characteristics. For example, past settlement, garden, or livestock *kraal* sites may produce long-term effects as a result of the sustained build-up of organic matter and soil nutrients which remain apparent many years after the abandonment of such areas.

Socio-economic factors also contribute to this variation within sites. This has been captured in this study by attention to between-farmer differences. In all study sites variants of wealth ranking were carried out to differentiate farmer categories according to local critieria.

In Ethiopia and Zimbabwe farm households were differentiated according to indicators of wealth defined by local informants. In Mali, a slightly different approach was used which focused on differentiating between soil resource management capability, interest and experience. Thus, in all study sites, detailed farm and field monitoring was carried out across wealth and resource management groups. While the sample sizes were necessarily small because of the intensity of data collection, the results do reveal some important patterns.

Not surprisingly, because of different access to resources – land, labour, capital and so on – different farmers manage their land in often quite different ways. But the case studies show that patterns of soil improvement and decline are not neatly correlated with wealth and asset status. Indeed, some of the well-endowed farmers showed the highest levels of nutrient depletion in their soils, in part because of their ability to achieve high yields. And, in fact, some of the lesser-endowed farmers were the ones who invested considerable amounts of labour in improving soil-fertility, and so yields, on their relatively smaller plots of land. A simple pattern of poverty-induced environmental degradation is not shown. Nor, indeed, is the opposite. The conditions for successful soil-fertility management at the farm level are multiple and interacting, just as at the more aggregate site level.

In exploring the great diversity of soil-fertility management strategies employed by different farmers across wealth and resource management groups, a number of broad 'pathways' of change can be identified (see below). These emerge from situating an understanding of soil management on farmers' fields in a historical context. By tracing the history of both fields and farm families, it is possible to see how the possibilities for effective and sustained soil improvement wax and wane with the fortunes of households and the influence of external events. For example, in the Ethiopia study sites the expansion and contraction of the garden area (*darkoa*) is dependent on the ability to mobilize sufficient manure and labour. This is seen to change over the demographic cycle as labour availability changes, in line with disease incidence, both human and animal, and in relation to cattle ownership, borrowing and sharing arrangements (see Chapter 2).

A pattern found across study sites is the differentiation between homefields and outfields.[23] Intensive styles of gardening, focused on organic matter improvement and often based on hoe cultivation, mounding and ridging, are found closer to the home in relatively small plots where productivity has been boosted through many years of investment. Further away, bush fields or outfields can be found which receive considerably less attention and show relatively lower levels of productivity and higher levels of nutrient depletion, unless given a boost through the addition of inorganic fertilizer.

Different farmers are able to pursue combinations of homefield and outfield cultivation in different ways depending on their asset base.

Differences often have a gender dimension, with men and women allocating effort to different areas of the farm; most often with women more engaged in intensive gardening efforts closer to the home, while men concentrate on the outfields.[24] While the survey elements of the study focused on the household as a unit of analysis, close attention was also paid to both intra and inter-household relations. As the case study chapters show, gender, age and status differences within households affect who does what in relation to soil management. Similarly, relationships between households are particularly important in influencing access to labour and cattle through cooperative loaning and sharing arrangements.

Emerging questions

The comparative approach across different scales – from country to site to farm to plot – has allowed this study to focus explicitly on diversity and dynamics and avoid the dangers of aggregation seen in the generalized policy statements highlighted above. In so doing the study has asked the following questions.

- What factors result in soil-fertility improvement or decline?
- What pathways of change are evident and how are these linked to broader livelihood strategies?
- What institutional and policy factors are important to encourage more sustainable soil-fertility management strategies in different settings?

Before highlighting some of the broad conclusions emerging from the study, it is necessary to lay out in some more detail the methodological stance adopted in this work, and how this complements but also, in some important ways, differs, from how soils have conventionally been looked at in Africa.[25]

DIVERSITY AND DYNAMICS: NEW PERSPECTIVES ON SOIL MANAGEMENT

As we have seen, the approach adopted by this study has emphasized the interaction of diversity and dynamics in soil management processes across a range of spatial and temporal scales. Interactions across scales – from micro-level soil processes to broader-level climatic and landscape changes – and in relation to different rates of change, are essential to an understanding of complex agro-ecosystems (see Allen and Starr, 1982; O'Neill et al, 1986; Swift, 1998; Noordwijk, 1999). This requires an integrated insight into both biophysical features and socio-economic processes, set in historical context. The following sections, then, highlight some of the aspects of both spatial diversity and temporal dynamics observed in the case study sites, before turning to a discussion of some of the methodological implications of this approach.

Spatial diversity

Spatial diversity is a key feature of soils in each of the sites. Local classifications of soils revealed the wide variation of soil types in a particular location, for example by mapping exercises with farmers (see examples in Chapters 2 to 4). While these may all belong to one soil series, the differences – across slopes, between areas with more or less erosion, under trees, or near termite mounds – have profound implications for the way farmers view and, in turn, manage soils. It is the management of such heterogeneity that is at the heart of farmers' own practice (Carter and Murwira, 1995; Brouwer and Bouma, 1997).

Thus soil niches, part of a complex mosaic of micro-variability in a farm, field or plot, may be critical to overall soil management. A relatively small area of high-fertility soil may be a critical resource within the whole farm, providing proportionately higher yields than other areas. For example, in Ethiopia the *darkoa* garden plot, created by continuous and long-term investment of manure and other organic matter, on average produces around double the yield of maize compared to the neighbouring outfield. For some farmers, crop outputs from the *darkoa* area amount to a significant proportion of the total contribution to household food supplies, despite the small area (see Chapter 2).

At the wider landscape level, spatial interactions between cultivated plots and biomass resources available in other areas are key. The harvesting of biomass – whether the collection of leaf litter or the transfer of nutrients through the manure of grazing livestock – is of vital importance in most of the case study sites. In some cases (for example in the highland Ethiopia site), the availability of extensive common grazing land is limited, and grazing must occur on more spatially concentrated areas: along roadsides, on the edges of fields, and in private grazing plots near homesteads. In other cases, extensive common grazing remains available, and the transfer of nutrients from grazing land to cultivated fields via livestock manure is central to soil-fertility management in the arable areas. Thus, the spatial patterning and availability of biomass resources at a broader landscape scale is critical to understanding the sustainability of the system (Fresco and Kroonenberg, 1992).

Given this spatial diversity at field, farm and landscape levels, mechanisms of positive feedback may result in continuous reinvestment in soil-fertility for particular sites, creating permanent 'hot spots' of high soil-fertility. Sites such as old cattle pens, settlement sites or gardens may attract livestock, for example, in the dry season because of the higher amounts and nutritional quality of grass, weeds or crop residues. The deposition of dung and urine at this time may again continue the process of fertility build up in such patches, with persistent effects over long periods of time (Blackmore et al, 1990).

While spatial diversity in soil properties is in part a result of biophysical parameters (underlying geology, soil type, topography, patterns of deposition and loss, etc), these have to be seen in a wider context in order to understand the changing patterns of diversity observed. The nutrient content of a soil, its pH and its cation exchange capacity, are in most instances the result of a complex interaction of biophysical and socio-economic processes over time.

Different social actors influence such changes in different ways. For example, richer farmers in the various case study sites are able to apply inorganic fertilizers and manure to some of their fields, thus enriching certain parts of the soil resource on their farm. Poorer farmers, by contrast, may adopt different soil-fertility management strategies, focusing limited fertility resources and investing more in efficient placement and timing in order to maximize returns, rather than adopting blanket applications. Wealth and asset status may not be the only factor influencing between-household differences in practice. For example, ethnic differences may be important where different groups adopt different soil-management styles based on long-established practices, perhaps developed elsewhere. Thus in Mali, in-migrants to the southern sites carry with them practices developed in the more arid zones of the north. Established forms of hierarchy and social position within an area may also imply differences in farming practice. Thus in southern Ethiopia, the remnants of an earlier caste system, as well as the past experience of landlord–tenant divisions, may result in old forms of soil management persisting as part of current practice.

Social relations within households may equally affect the nature of the soil resource. For example, gendered cropping styles and practices may result in different types of cultivation practice, choice of crop and use of fertility resources in different sites. For example, in Zimbabwe women are particularly engaged in the gardening of vegetables, which involves particular types of mounding and ridging techniques and the incorporation of organic matter to create a rich soil resource. Similarly, in Mali changing forms of domestic organization with the frequent break-up of large patriarchal households into smaller, more nuclear units has major impacts on the way labour is organised, and the nature of obligations towards the management of communal family fields by women and junior men. The result of such socio-economic differences and social relations is inevitably a different patterning of soil-fertility on each farm. Such diversity may arise also as a result of the unintended consequences of other actions – for example the location of settlement or a livestock *kraal* may result in increased concentrations of fertility resources in particular places which can subsequently be made use of for agricultural production.

Temporal dynamics

Superimposed on such spatial diversity at different scales are issues of temporal dynamics operating at different rates and over different time scales. Within each season, changes in nutrient availability are the result of changes in mineralization and decomposition rates prompted by changes in rainfall and microbial activity. Thus early season flushes of nutrients may be important, and require strategies for their capture, such as early or dry planting. Between-year variations are also significant, with higher and lower rainfall periods resulting in different levels of available soil moisture and nutrients. Over longer periods, processes of mineralization and immobilization may also affect the availability of nutrients in mobile or immobile pools (Woomer and

Swift, 1994). In addition, different elements of the nutrient cycle change at different rates, with some mineral elements (notably nitrogen) showing much greater variability over time than others (such as phosphorous). Thus assessments of simple aggregate availability of fertility resources may be insufficient to assist with the complex task of synchronizing highly temporally-variable nutrient availability with plant growth (Woomer et al, 1994). In the annual farming 'performance' (see Richards, 1989), timing is all, requiring skilled insights into soil–crop interactions and the dynamics of change.

Of major concern to farmers and policy-makers alike is the question of whether soil-fertility is declining or improving. In answering this we are concerned with somewhat longer trends over time. If we are to make any statement about change we must be able to detect trends in data against a background of variability (cyclical or simply 'noise'). We must also be sure that the trend we are seeing is a real one, not driven by another variable. For example, in Wolayta, Ethiopia (see Chapter 2), maize yields are influenced by the both rainfall and the availability of fertilizer. No trend in yield potential could be confirmed over the period from 1971–1993, as fertilizer use increased and then decreased, with yields returning to their pre-fertilizer levels. No overall trend in rainfall was seen during the period, suggesting that this was not a confounding variable. Therefore the study concluded that the data could not be used as evidence for yield reduction due to soil-fertility decline (Eyasu and Scoones, 1999).

Another key element of detecting trends is to be sure about what the baseline is. In the Ethiopian example, the baseline yield level was that before the widespread application of fertilizer. However, if the starting data point used was at the peak of fertilizer use, then a declining, rather than cyclical, yield trend would be detected leading to possibly quite different conclusions. When monitoring soil parameters, baselines are always critical but often quite difficult to define, due to seasonal and interannual variation. A final important question to ask is: what is the indicator of change which is of most interest? As discussed earlier, it may not be appropriate simply to use technical measures of soils to assess degradation, for example, unless such parameters are directly linked to wider values for livelihoods. Thus, choosing a set of indicators that link soil-change processes to livelihood values is a critical step. Too often indicators of land degradation or land quality[26] appear to be plucked out of a hat and do not pose the question: does change in the particular indicators chosen really matter?

Changes in soils may not be associated with smooth, secular trends, but often with relatively sudden transitions between states. Environmental transformation therefore may be more reliant on contingent and chance events, than predictable, slow evolutionary change. Methods for identifying such transitions, and the key events and conditions surrounding them, are therefore vital (see below). This requires taking a historical perspective which locates soil transitions in various time periods (over years, decades, centuries or even millenia). The conjuncture of particular combinations of events is often key to such explanations, and may require a retracing of ecological, economic and social histories as part of the investigation. In developing such historical insights for particular plots, farms or landscapes it is important not to infer historical trends from spatial patterns.

Where major interventions have occurred which have fundamentally reshaped land use and agricultural practices, a historical imprint may be left on the landscape. For example, the legacies of technocratic planners can be detected in the soils of Ethiopia as a result of the imposed villagization schemes of the 1980s (Chapter 2), in Mali in the form of organization of the large rice irrigation schemes, and in Zimbabwe resulting from the land reorganization imposed during the centralization and land husbandry periods. Thus soils in each of these cases are, in part, the product of past interventions, some dating back over 50 years. Past practices may leave both positive and negative legacies which influence current options. For example, earlier settlement or *kraal* sites are widely valued particularly for new garden land (see Chapters 2 and 4), where the regular deposition of household waste, excrement and dung has resulted in the concentrated accumulation of nutrients.

Agricultural landscapes are thus made up of a mosaic of high and low fertility sites, each with distinct dynamic histories. Each, in turn, requires different management strategies. The result is the need for site-specific approaches to soil-fertility management that take note of such diversity and changing soil patterns, and build on the adaptive, responsive 'performance' of farmers' cultivation strategies. As discussed in detail in each of the case study chapters which follow, farmers are well aware of such challenges. The application of soil amendments, for instance, is often highly focused both in space and time, with placements being made to improve particular patches or capture particular moments when nutrient-uptake efficiencies are maximized. In such diverse and dynamic settings, then, surprise, uncertainty and variability are the norm. This requires highly dynamic soil-fertility management approaches that are at once, opportunistic, efficient and flexible.

Integrating understandings of natural and social processes

As with the analysis of spatial diversity, insights into temporal dynamics must take into account the range of socio-economic influences driving change. Understanding how soils change, and what the challenges for soils management are, therefore requires insights into the histories of landscapes, fields and plots. Histories of clearance, cultivation, settlement, burning, grazing and planting are intimately connected with the social, economic and political histories of human action. Thus particular types of farming practice may be redolent with social meaning and identity and so imply forms of validation for particular social arrangements (see Guyer, 1984). An integrated understanding of the natural and social worlds is therefore required if the observed diversity of soils is to be interpreted with any success.

Soils can therefore be seen as both a template for and a product of social action. Social relations, domestic organization, labour practices, forms of hierarchy and social position all impinge on the 'social life' of soils (see Nyerges, 1997, after Appadurai, 1991). The way individual farmers influence soils is mediated by a range of formal and informal institutions. Thus the way input markets function affects the degree to which inorganic fertilizers are used as a soil amendment, for instance. Similarly, institutions governing land holding and tenure may affect the degree to which farmers invest in soil improvement, particularly for the long term. Levels of available labour, governed by both inter- and intra-household gender and other social relations, may have big impacts on the way soil-fertility management is organized, where soil-fertility investments are made, and what is applied. Institutions affecting access to credit or savings may also have an impact on the ways soils are managed, by affecting who has access to cash and when (see Chapter 6). For example, in southern Ethiopia a whole range of local institutions exist which facilitate access to labour, draft oxen, credit and other means of production. Investing in the social relations and networks associated with these is a critical means of survival, especially for the poor (Berry, 1989). Yet institutions affecting soil management and farming practice are not stable – continuous renegotiation at the local level, resulting from shifting political and social relations, makes for a great deal of flexibility and fluidity. When linked into wider circuits of economic change, education, development activity or migration, the interaction between local institutional forms and wider contexts becomes key (Berry, 1993). Thus, as seen in the Zimbabwe case study (Chapter 4), changes in economic policy in the early 1990s have had a major impact on soil-management practices, filtered through the changes in social institutions (particularly gender relations surrounding land and labour) at the local level.

Understanding the complex dynamics of soil transformation thus requires an integrated insight into spatial and temporal dimensions across a range of scales, and integrating not only a range of natural science perspectives, but also, crucially, an understanding of social, economic and institutional processes. The frameworks and methodologies necessary for gaining insights into such complexity are the subject of the next section.

Understanding the complex dynamics of soil transformations

An appreciation of diversity and dynamics suggests a set of questions and methodological challenges which, while not necessarily new, are not often asked in conventional studies of soils in Africa. Box 1.1 offers a checklist of some of the key questions which were addressed as part of the studies reported in this book. Others could be added, and other combinations explored.

Posing questions of this sort pushes us to think about methods for answering such concerns. Again this perspective emphasizes an interdisciplinary approach to enquiry, one that integrates methods which adopt a 'hybrid' approach (see Batterbury et al, 1997) to the investigation of environmental and agricultural issues. Box 1.2 highlights the wide range of methods used in the case studies reported in Chapters 2 to 4.

Because of the large number of variables, the complex spatial patterning of soils, and the multiple time dimensions over which soil processes operate, non-linear dynamics are almost inevitable. Uncertainty and surprise are always key features in such situations (Holling, 1993). Thus methods for identifying key driving variables, important transitions, and system boundaries and discontinuities are required. So how is it possible to make sense of all this

BOX 1.1 UNDERSTANDING CHANGE: SOME KEY QUESTIONS TO ASK

Spatial diversity

- At what scales should measurements be taken?
- What are the spatial units identified as important by farmers?
- What criteria differentiate different spatial units?
- What is the historical origin of current spatial patterning?
- How should insights derived from different scales be related to each other?

Temporal dynamics

- Against what baseline should change be assessed? What are the key indicators of change? What factors make a difference?
- What are the longer-term dynamics of the system? Is observed change a temporary hike, part of a cycle or the consequence of a longer-term shift?
- What significant thresholds exist for both soil improvement and degradation processes?
- What endogenous and exogenous factors influence changes in soil-fertility?
- In the past what combination of factors and events have resulted in major shifts?

Source: adapted from Scoones and Toulmin, 1999

BOX 1.2 SOME METHODS FOR UNDERSTANDING TEMPORAL AND SPATIAL DYNAMICS

Spatial diversity

- Local terminology and classification of states
- Mapping of soil types by farmers
- Landscape and site histories
- Resource-flow models by farmers
- Partial nutrient budgets
- Farmers' experiments with spatially-differentiated treatments

Temporal dynamics

- Local terminology and classification of soil transitions
- Archival records and travellers' reports
- Biographies and life histories
- Oral histories of environmental change
- Field and site histories
- Aerial photographs and satellite images
- Time-series census and experimental data
- Natural experiments with long-term 'treatments'

Source: Scoones, 1997

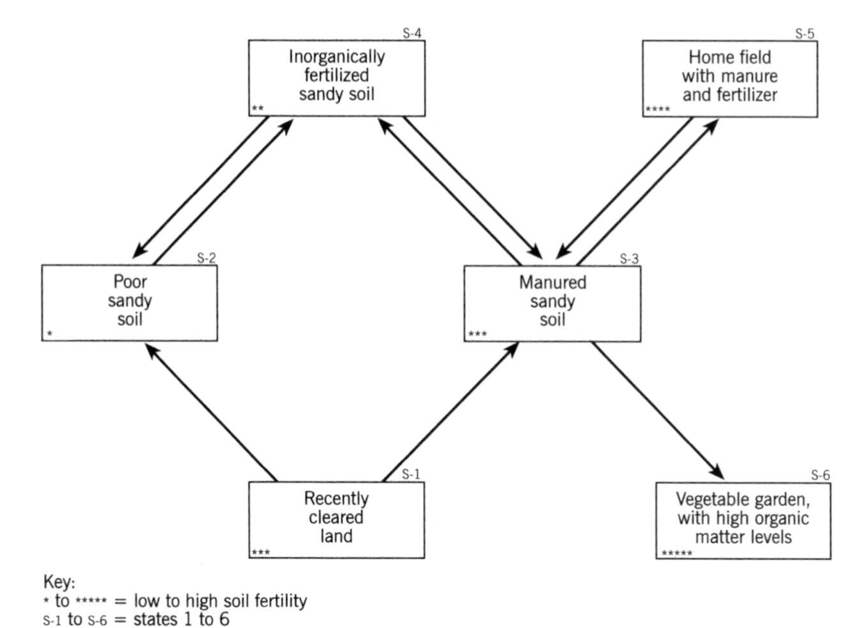

Figure 1.4 *Dynamics and diversity in Zimbabwean farming systems: a 'state and transition' model*

complexity? What frameworks for analysis and intervention make sense? A 'diversity and dynamics' approach suggests that, rather than a linear view of soil change, multiple possible states should be envisaged, characterized by distinct physical, chemical and other features. The stability of any of these states will depend on a variety of factors, both biophysical and socio-economic, which drive the transitions between states and affect their frequency. For example, there may be cases where only one single dominant state is found, where rainfall conditions, burning, cultivation, grazing, dung deposition and other factors remain constant. This, however, is very rare, and most situations are highly dynamic.

Figure 1.4 presents a 'state and transition' analysis[27] from Zimbabwe. Here six different 'states' are identified for the sandy agricultural soils in Chivi communal area, with a series of 'transitions' between them. Retracing patterns of change over time for particular sites highlights how, in any one site, all identified states can exist in an area, both sequentially and in parallel, depending on the factors influencing the various transitions. For example, on clearance from *miombo* woodland (state 1) and the creation of agricultural fields, many soils lose fertility over time and a low-level equilibrium soil-fertility level is reached (state 2). Several paths are possible from this point. Either the land is left to continue to produce at a low level (state 2), or different types of investment are made. Between the 1950s and 1970s, many farmers were able to add manure, other organic matter or fertilizer to such soils (states 3 and 4) in order to boost fertility, although this relied on the availability of cattle, labour and cash used to buy fertilizer. With declining cattle populations due to drought, and the rising cost of fertilizers, from the late 1980s onwards many farmers have let their outfields return to state 2. Instead, a more selective investment of labour and organic matter has been focused on smaller homefield or garden sites (states 5 and 6). Thus, over time, a variety of biophysical factors (eg rainfall) and socio-economic variables (eg labour, tenure, etc) influence the patterning of soil resources through both sudden key events (eg drought, fertilizer price rises, etc) or slower changes (eg labour availability through changing patterns of migration). A number of important steps need to be included in such an analysis (Box 1.3).

This approach starts with an understanding of the local situation; insights into the differentiated agro-ecological, socio-economic, institutional and policy contexts are an essential starting point. This requires a participatory approach to investigation, which draws from farmers' own understanding of their situation, and the changes that have occurred in the recent past. Such local level, participatory analyses may be linked to more conventional research on key aspects, but the questions to be asked – whether by natural or social scientists – must derive from understanding the local setting. The perspectives that emerge can provide important insights for future action, whether in terms of field-level action research on particular technological or management options, or institutional or policy interventions which, in different ways, also encourage particular transitions to desirable states. In order to encourage such analysis, and links to practical action at different levels, tools are required which allow communication and joint analysis – by farmers, researchers, extension workers and others. The simple approaches of resource mapping, option ranking, flow diagrams and so on, described in the case studies and reviewed in more detail in Chapter 5, all provide ways in which a focus for analysis and common understanding can emerge, which links analysis to action, and allows collaborative approaches to intervention and monitoring to emerge.

SOILS, AGRICULTURE AND LIVELIHOODS: MULTIPLE PATHWAYS OF CHANGE

As discussed above – and as the case study chapters show in more depth – there exist multiple pathways of agricultural and environmental change, both between and within sites. In some cases land is being actively improved, while in others the soil resource is losing nutrients and productivity. Understanding these changes requires locating such patterns of soil improvement or degradation in a wider context. It is therefore necessary to ask: what are the interactions of factors which influence soil change?

A number of different pathways of environmental and livelihood change are offered in the literature and the policy commentaries emerging from these debates. As discussed earlier, much of the mainstream policy narrative on soil change in Africa is based on a neo-Malthusian interpretation of the interaction between population and environment. A 'downward spiral' of environmental degradation and poverty is, it is argued, the inevitable result of increasing population pressures (Cleaver and Schreiber, 1995). While this may perhaps represent the dominant popular interpretation, some alternatives are

BOX 1.3 KEY STEPS IN A DIVERSITY AND DYNAMICS ANALYSIS

- The range of possible states are identified and their characteristics defined. This may require soil mapping at various scales – landscape, farm, plot. Building on local classifications of both 'states' (soil types) and 'transitions' is important.
- The common transitions are noted – those that result in both positive and negative change (as defined by farmers' own objectives). The factors that influence these, including both biophysical and social/institutional processes, are then identified. Simple flow diagrams (as in Figure 1.4) can be constructed to highlight options.
- A key part of this analysis is to see how different factors have combined in the past. This requires the compilation of an event history for the site and highlighting key events and conjunctures over time. A simple timeline derived from key informant interviews can assist in developing an understanding of how key events combine and influence patterns of environmental change.
- The range of desired states is identified (for different groups of people) through discussion. The transitions required to increase the likelihood of such states are then identified (from the flow diagram).
- The feasibility of effecting different types of transition for different groups of people then can be assessed in relation to their existing access to key assets (eg in relation to the availability and access to natural, social, human, physical and social capital).[28]
- Trade-offs in outcomes are then assessed (eg immediate yield increases through fertilizer application versus long-term investment in sustainability through organic matter applications) and priorities established with farmers (which will vary by farmer and type of plot or crop focus).
- Institutional and policy constraints to achieving the desired outcomes are also assessed, with other types of institutional, organizational and policy intervention identified (eg in relation to specific areas of technology development, credit support, tenure reform etc – see Chapter 6).
- Starting at the local level a process of action planning, monitoring and learning can be initiated, focusing on what is possible given existing patterns of access to assets and existing institutional and policy constraints. Simple innovations based on local experimentation and monitoring may highlight further challenges (see Chapter 5).

also suggested which counter this perspective (Forsyth et al, 1999). These argue that, while environmental degradation and soil-fertility decline are certainly problems, we must be careful in putting forward generalized statements. Indeed, a more differentiated look at particular situations shows the possibilities, under particular circumstances, of improvements in environmental conditions associated with the intensification of agriculture and the reduction of poverty.[29] Under this argument, increasing population density changes the incentives to invest in land, resulting in labour-intensive processes of agricultural and environmental improvement (Boserup, 1965). This is particularly apparent when a range of policy conditions are assured, including access to markets, good quality infrastructure, knowledge and technology and secure tenure (Tiffen et al, 1994).

Table 1.3 *Relationships between key contextual variables and soil-fertility management practices and outcomes across sites*

	Ethiopia		*Zimbabwe*		*Mali*			
	Highland	*Lowland*	*M'wende*	*Chivi*	*Dilaba*	*Siguiné*	*Tissana*	*M'Péresso*
Rainfall (mm/year)	1272	924	850	550	450	450	650	800
Soil type	Volcanic nitosols	Volcanic nitosols	Granite sands	Granite sands	Lithosols, acrisols	Lithosols, acrisols	Lithosols, acrisols	Lithosols, acrisols, gleysols
Infrastructure	**	*	****	**	***	**	**	***
Markets	**	*	****	**	***	**	** (for rice)	*** (for cotton)
Extension services	**	**	***	**	*	*	****	****
Land tenure security	***	**	****	****	****	****	***	****
Population density (people/km^2)	375	110	150	44	50	15	29	18
Fertilizer purchase (%)	81	87	66	16	100	0	0	100
Cattle ownership (n)	5	4	5	4	13	22	26	19
NUTBAL (N)	–64	–30	–17	–28	–24	–32	+34	+35

Notes: For Ethiopia and Zimbabwe, these figures are based on averages for maize fields across all resource groups for outfields. For the Mali cases these relate to rice and cotton for Tissana and M'Péresso, and millet fields for the other dryland sites.

Key: **** = more to * = less

What scenarios are evident across the case studies, and are there particular pathways of environmental and livelihood change which can be seen? Table 1.3 pulls together some general data on each of the research sites and relates these to a set of indicators of soil change – manure inputs (with cattle ownership as the proxy indicator), fertilizer use and nutrient-balance estimates on main fields. At this aggregate level it is difficult to discern consistent and distinct patterns. The data show that at a site level, there are no simple correlations between soil-fertility management practices, nitrogen balances and factors such as rainfall, population density, market access, infrastructure, extension coverage and tenure security.

Clearly a combination of factors affect outcomes. Thus simple arguments based on biophysical characteristics or on demographic factors (in either a Malthusian or simple Boserupian form), do not hold, as a more integrated analysis is required. As Sara Berry (1993, p183) argues:

> *Agricultural intensification ... cannot be reduced to a question of change in relative factor proportions. Instead, changes in agricultural technology must be understood in relation to changes in the organization of agricultural production and specific regional configurations of economic, political and social change.*

The case study chapters that follow attempt to take such a broad and historically situated perspective in analysing changes in soil-management practice across the sites. Overall, though, three broad groupings of sites can be identified, based on different pathways of change, each with different implications for patterns of intensification and sustainability.

Dryland farming: opportunistic cropping on low-fertility soils

These sites are characterized by low agricultural potential, largely due to low levels of rainfall. The relatively extensive dryland sites such as Siguiné and Dilaba in Mali and Chivi in Zimbabwe could be described in this way. All have relatively poor soils, receive low annual rainfall levels and suffer periodic droughts. The limited inherent potential of these areas means that investment in soil improvement has low and uncertain returns. Overall, the result is a relatively low-input and low-output system. Due to the huge interannual and seasonal variability in rainfall, farmers must be highly responsive in their farming approach. If good rains are received, then it may be worthwhile investing in labour and soil-fertility inputs, whereas in many years this does not pay. An opportunistic approach to farming is the result, reliant on careful agronomic responses to an unfolding season (Scoones et al, 1996). In some years, through the efficient timing and placement of fertility inputs, significant yield responses can be achieved if the broader conditions, outside farmers' control, are right. In other years fields are often left largely alone following planting, and what yield is achieved is regarded as a bonus.

Nutrient budget data from case study sites of this sort show net nutrient losses due to the low level of inputs applied. However two caveats must be applied to this data. First, such losses vary considerably from year to year because of the high variability of yield levels, which means that single-year data should not be taken too seriously as a guide. Second, because inputs in such areas are often applied in a highly spatially-focused manner, fertility levels in the plant zone may be high with good uptake efficiencies resulting, while the surrounding soil may have very low fertility levels. In the Zimbabwe case, for instance, the poor granite sands of many outfields act more as a planting substrate, with limited and focused applications creating a response. Aggregated pictures, even at a plot level, therefore, may not reveal the complex spatial dynamics of soil-fertility management in such areas where, despite very low inputs levels in total, high responses can be achieved (a low input–high output system) during intermittent good conditions.

Overall, though, net nutrient depletion seems to be occurring. But does this matter? The soils of southern Zimbabwe and the Sahel have long been very low in nutrients and yet crops are still grown. In the Sahel, for instance, inputs from Harmattan dust and mineralization of the limited available organic matter may result in around 15kg/N per hectare per year – an amount approaching the level of extraction noted in the partial nutrient balances. As long as some fallowing occurs to regenerate a limited amount of organic matter in the soil, this may be a reasonably sustainable farming system. Indeed, continued depletion may not matter hugely as long as responses to focused applications can be achieved, and the efficiency of input use is continuously improved. This may be possible as long as organic matter levels do not drop below a lower threshold of around 1 per cent (see Pieri, 1989).

Declining land quality and agricultural involution

A second cluster of sites can be found in areas of higher potential, but where agricultural productivity and soil-fertility is stagnant or declining. The outfield sites in Ethiopia (particularly the highland areas) and parts of Mangwende in Zimbabwe could be described in this way.

Despite increasing land pressure, a pattern of agricultural intensification and associated investment in the soil resource is not observed to any significant extent. Instead of a Boserupian cycle of improvement, a more negative picture of agricultural involution is observed (see Geertz, 1968). Under such circumstances a low asset base, combined with an unsupportive policy environment, create conditions of limited productivity. This, in turn, results in reduced capacity and few incentives to invest in soils leading to yet further declining productivity. In essence, this is the 'downward spiral' which is so dominant in mainstream policy narratives on soil and land management in Africa (see above).

A range of factors noted across the sites may contribute to such a pattern of low investment. In all sites in this cluster, significant potential exists to boost productivity through the application of fertility inputs. In contrast to the drier sites, relatively reliable and high rainfall and, in the case of Ethiopia, relatively good soils, mean that the addition of manure or fertilizer (or some combination) can result in reasonable yield increases. Such potentials are demonstrated in numerous research trials and have been witnessed in periods when fertilizers have been subsidized and supplied effectively (as in the period when the WADU integrated rural development project operated in Ethiopia and the period after independence in Mangwende, Zimbabwe; see Chapters 2 and 4). However, small plot sizes mean that livelihoods must be sustained through means that go beyond the intensification of agricultural production, particularly in Ethiopia (see Chapter 2).

A combination of factors have made such investment options limited. Changes in pricing and marketing arrangements following structural adjustment and agricultural liberalization have seriously affected the profitability of fertilizer use in all case study countries (Chapter 6). Combined with poor infrastructure and credit facilities in many settings, this means that, for many (but not all – see below), the ability to use fertilizer inputs to boost productivity levels has been limited. Other options, based on organic sources, have similiarly been constrained by the lack of available cattle, biomass or labour.

In such situations, nutrient depletion is highly evident, with potentially longer-term consequences. Since this type of pathway is mostly associated with asset-poor farmers in these sites, this trend has serious consequences for poverty and livelihood sustainability.

Agricultural intensification and soil investment

A final cluster of sites (or parts of sites) are associated with a much more positive process of agricultural intensification and investment in soil-fertility. Nutrient-budget analysis shows how such areas may show patterns of nutrient accumulation or at least stability (details in Chapters 2 to 4). A number of different types of intensification can be observed.

- First, there are the labour-intensive gardening systems in homefields and gardens found in all sites, where the application of organic matter (manure, compost, household waste, leaf litter etc) results in sustained improvements in soil-fertility levels and yields. Such systems rely on larger areas – usually of common land – from which such organic inputs are harvested. Of particular importance are interactions between the cropping and livestock elements of the farming system, with livestock assisting in the concentration of nutrients in the garden areas through the production of manure.
- Second are niche-focused strategies based on particular sites such as valley bottomlands or river banks which have inherently higher levels of soil moisture and nutrients due to their position in the landscape. These areas are often sites of highly intensive production, where a wide range of higher-value crops may be grown. In all sites, intensification of agriculture in particular landscape niches, making use of the varied toposequences found, are important features of soil management practices. In many cases such low-lying sites require more labour to cultivate, and therefore such strategies tend to emerge only when other sites show lowered productivity, or land pressures result in people seeking alternative, more costly options (Scoones, 1991).
- Third, there are areas where farmers have managed to invest in inputs for their main fields. Richer farmers, with access to cash from various sources, in all study sites are observed to apply fertilizer and manure to their outfields and manage to improve yields to reasonable levels, especially if the rainfall is reliable. In areas such as the cotton zone of Mali, where there is significant infrastructural and extension support for particular crops, then soil-fertility investment may occur across a broader group of farmers. This is particularly reliant on credit systems which allow poorer farmers to purchase fertilizers and livestock to provide manure. In the Ethiopia case studies, particularly in the lowland site, the current credit package focused on improved maize, and fertilizers has not managed to reach beyond the relatively asset rich group (see Chapter 2), which limits agricultural intensification.
- Fourth are areas where investments in irrigation have reduced risks associated with water limitation, making investment in soil amendments

Table 1.4 *Cases of soil-fertility improvement*

	Description of changes	*Nutrient balances (kg/ha)*	*Key assets influencing*	*Key external drivers influencing*
Ethiopia – *darkoa* gardening	High inputs of manure and other organic material, combined with intensive hoe cultivation (mounding, ridging). Long-term accumulation of soil organic matter and nutrients. Highly productive	Richer resource group, garden areas: Enset N: +11.5 P: +11 Taro N: +4 P: +10.5	Female and child labour for collecting and transporting; cattle ownership/holding for manure	Drought and disease affecting cattle populations
Mali – irrigated rice zone	Highly productive rice farming; fertilizer applied in excess of recommended rates. However K removals from livestock grazing of stubble is significant	Average all resource groups (rice field) N: +34 P: +8 K: –88	Good fertilizer supply; support from Office du Niger; herding and manuring arrangements with Fulani herders	High value cash crops (rice and vegetables), with good marketing opportunities
Mali – fertilizers on cotton	Recommended inputs of fertilizer supplied on credit as part of CMDT package results in high cotton yields in reasonable rainfall years	Average all resource groups (cotton fields) N: +35 P: +2 K: +5	Cash for purchase of fertilizer and repayment of credit	Fertilizer price and credit system
Zimbabwe – mixed manuring and fertilizer placement	Careful timing and placement of manure and fertilizer (in planting hole or in furrow) results in significant yields of maize in good rainfall years with relatively low input levels	Richer resource groups, homefield Mangwende N: +51 P: +16 Chivi N: –13 P: +12	Cash for purchase of fertilizer; cattle for manure; timely draft power, labour for placement; skill and knowledge for placement	Fertilizer prices and markets; drought and cattle availability

advantageous. The rice zone site in Mali is a good example, where the combination of irrigation infrastructure and support (in terms of extension advice, credit support etc) through the Office du Niger has resulted in increasing investments in soil management.

Two key questions follow from this analysis. First, what factors influence transitions between these pathways for different people in different sites? Second, in what ways can external influences encourage positive transitions and help prevent negative ones? To begin to answer these, we need to examine the dynamics of change in a bit more detail. This requires looking at patterns within the study areas, differentiating by both people and place. Table 1.4 presents a series of case examples from across the three case study countries of positive change. This describes the type of changes occurring, the current level of nutrient balance, and the key factors influencing such outcomes. The table differentiates between access to assets at a local level (including land, labour, draft power, skills, social resources) and external drivers influencing change.

Within sites, then, we see a range of factors influencing why, on a particular piece of land, soil-fertility may improve or decline. In terms of soil improvement strategies, two broad patterns can be identified (Carswell et al, 2000).

First, a labour-intensive approach based on manuring and the application of other organic matter to a relatively small area. Such a gardening style of agriculture is common across the sites, particularly in homefields, and represents an important way in which soils are enriched and transformed. This requires high levels of available biomass (eg leaf litter, compost), manure or other organic waste, as well as considerable labour for composting, carrying materials to the fields, and for the labour-intensive styles of cultivation often associated with gardening (eg ridging, mounding etc). Given that such investments often take many years, with effects cumulative over time as the level of soil-fertility and productivity increases, a level of tenure security is critical. As the villagization experience in Ethiopia showed (see Chapter 2), the forced abandonment of such resources can have significant impacts on livelihoods.

Second, a more capital-reliant approach is observed, based on the purchase of inorganic fertilizer. Here external factors are critical, including price ratios, input markets, and infrastructure (including, in the case of the rice zone in Mali, the maintenance of the irrigation system). The case of the cotton zone in Mali, where the parastatal CMDT has provided a range of support, is perhaps the most capital-focused example. However, reliance on such factors may prove risky, as the experiences of a number of sites have shown during the structural adjustment period.[30] Where input prices increase dramatically relative to output prices, alternative strategies may emerge which combine the addition of inorganic fertilizer with more labour-intensive fertilizer-placement strategies. In such situations (see, for example, the discussion in Chapter 4 on Zimbabwe), lower amounts of fertilizer are applied, but uptake efficiencies are improved.

But the story across the sites is not all positive, as a decline in soil-fertility is seen in some sites. A number of factors contribute to this. In some situations this is as a result of a conscious switching of investment to other parts of a farm (for example, garden areas in southern Zimbabwe or Ethiopia), with land left for more opportunistic cropping, and expected yields at a low level. In situations where ensuring the right balance of soil nutrients and available water is difficult, this may be an appropriate response to the inherent riskiness of dryland areas, as long as yields can be increased elsewhere. In other situations a decline in soil-fertility has a more direct impact on livelihoods. For example, declining yield levels of high value crops (such as maize in Mangwende, Zimbabwe or cotton in Mali) on main fields may have serious consequences if not offset by additional inputs. In such situations 'nutrient mining' may result in negative impacts on livelihoods and, in the longer term, if organic matter levels for instance drop below critical levels, for the sustainability of the system.

Such negative pathways of change may arise for those without sufficient assets at their disposal (cash, labour, oxen etc). This may be a result of declining state support (as in the case of structural adjustment and liberalization impacts), the impoverishment of the asset base through the impact of drought, the failure of markets to provide alternatives (eg through credit for fertilizer) or the inability to raise income through alternative off-farm sources and remittances for input purchase or asset rebuilding. Such factors clearly influence different people in different ways. As the case study chapters show, some end up in a vicious circle, where a declining asset base combines with institutional and policy impediments. Such patterns may result in a process of 'agricultural involution' where poverty increases as the resource base declines. However, as the case studies show, such a 'downward spiral' and a direct linkage between poverty and environmental decline is not universal. Indeed, many poorer farmers are able to intensify and improve soil-fertility at the same time along a pathway of labour-led agricultural intensification and resource conservation. By contrast, some richer farmers may fail to invest in their soils, preferring instead to maximize short-run returns or rely on other sources of income for their livelihoods.

The consequence, then, is a complex interaction between livelihoods, poverty and environmental change, with no predetermined outcome. In order to understand such dynamics, it is necessary to unpack the relationships between the broader context, the assets held by different households and individuals and outcomes, both in relation to changes in people's livelihoods and the resource base on which they are, at least in part, reliant. Processes of soil change therefore must be seen in a wider livelihood context, where influences ranging from macro-policy factors to micro-household-based factors all impinge, and are mediated by a complex interaction of institutions and organizations located across levels. In order to identify the range of pathways of livelihood change and their influence on soils for a particular site and grouping of people, a number of questions must be asked (see Box 1.4).

Thus pathways of environmental change – and the associated processes of environmental sustainability, land degradation and soil enrichment – are intimately connected to farmers' livelihood constraints and opportunities, as influenced by the broader setting, the available capital asset base, and the range of institutions and organizations mediating outcomes. In any particular case, then, a technical understanding of soils must be allied to a broader understanding of livelihood change if the underlying factors influencing the prospects for a more sustainable use of soils are to be grasped.

BOX 1.4 PATHWAYS OF CHANGE: LINKING LIVELIHOODS AND SOIL MANAGEMENT – SOME KEY QUESTIONS

- What different strategies for soil management are being employed by different people in different sites (eg different styles of agriculture – from organic gardening to cash cropping with fertilizers etc)?
- What are the consequences for people's livelihoods (eg changes in poverty levels, changes in degrees of vulnerability) and the resource base (eg changes in nutrient balances, levels of soil conservation etc)?
- What are the contextual factors influencing these different soil-management strategies? What broader trends are evident, and what shocks or risks are significant?
- What are the key assets necessary for sustainable soils management? How are these differentiated between sites and among different groups and individuals?
- What institutions and organizations affect the ability of different people to gain access to the necessary assets required for both improving livelihoods and sustainable soil management?

Such an analysis pushes us towards a more holistic assessment of the intervention possibilities and policy options for encouraging more sustainable livelihoods and soils management. Thus an analysis of contextual factors may identify some significant trends or risks amenable to external influence. In the Mali case, for example, changes in the fertilizer price and supply network through liberalization policies may have negative effects on soil management and livelihoods in the cotton zone, with many implications for the institutional and organizational questions surrounding support to such areas. Similarly, an assessment of the distribution of the asset base may reveal some key constraints. For example, the decline in cattle populations due to trypanosomiasis in the lowland areas of the Ethiopia case study area is having major consequences on the ability of farmers to pursue a manure-based intensification strategy on home fields. But it may not be material assets alone which constrain opportunities. Access to knowledge and skills and social arrangements for improving soils may be just as important, particularly for those whose material asset base is limited. Thus investment in a knowledge-focused participatory extension strategy with farmers, linked to the encouragement of farmer groups, may be appropriate in some settings (see Chapter 5 and Defoer et al, 1999, for further discussion of this theme). A focus on institutional and organizational factors may also highlight areas for concentrating support. For example, the supply of cheap fertilizer is a key issue for many farmers across the sites. Credit markets, infrastructure support and input and output marketing arrangements are all highlighted as critical constraints. Chapter 6 explores in more detail the range of possible intervention options and policy issues that arise from an examination of the case study experiences.

CONCLUSION

In contrast to the generalized statements that dominate the policy debate, the research discussed in this book points towards the need for a much more nuanced perspective. Such a perspective must take into account the spatial and temporal variations in soil properties and dynamics and link understanding of biophysical processes and socio-economic change. A historical perspective highlights the importance of looking at environmental and social change over the longer term. A range of influences push pathways of change in different directions. These are not continuous and predictable, as the interaction of biophysical events (such as drought) with changes in macro-economic policies (as with structural adjustment) lead to shifts in institutional configurations and so farming practices at the local level. The findings across the sites therefore echo those of Sara Berry (1993, p189) when she comments:

> *Agricultural intensification has neither been inevitable nor continuous in African farming systems. In some areas, intensification was halted or reversed by changing environmental or political and economic conditions; in others, it has occurred not as an adaptive response to population growth or commercialization, but in the face of growing labour shortages and declining commercial activity. Such cases underscore the importance of studying farming as a dynamic, social process.*

The challenge then is to find ways of improving the possibilities of successful soil-fertility management under smallholder conditions through an appropriate combination of policy and technical support (see Chapter 6 for a detailed discussion of this theme). A context-specific approach to the analysis of soil-fertility issues requires a different style of research. Instead of attempting aggregate analyses leading to broad plans of action and statements of policy, a more differentiated perspective is needed. This needs to build on local understandings of processes of change, and capitalize on opportunities for action identified at the local level. While many of the technologies and interventions conventionally recommended may remain appropriate, these need to be fitted to particular settings. In order to ensure that research and technology development are focused on local needs, rather than responding to a simplistic and generalized policy agenda, such work needs to be firmly linked to a participatory learning approach which encourages local-level innovation, testing and adaptation (see Chapter 5 for a further discussion of this theme).

Soil management which takes account of dynamics and diversity therefore requires an approach that links soils and people, integrating the technical and the social in both analysis and action. This requires new ways of thinking and acting that build on interdisciplinary perspectives and innovate with new styles of participatory research, action and learning. The case study chapters which follow demonstrate how such an approach might look in practice, while the concluding chapters reflect on the implications for field-level research, action and policy respectively. We hope this book will provoke new field-level activities, as well as encouraging reflection on the policy debate and the focus of development efforts in this important area.

Chapter 2

Creating Gardens: The Dynamics of Soil-fertility Management in Wolayta, Southern Ethiopia*

Alemayehu Konde, Data Dea, Ejigu Jonfa, Fanuel Folla, Ian Scoones, Kelsa Kena, Tesfaye Berhanu and Worku Tessema[1]

Introduction

Questions of soil management feature prominently in the policy debates on the future of Ethiopian agriculture and environment. Very often, a pessimistic picture is painted, with dramatic prognoses of environmental catastrophe. In particular, soil erosion has been highlighted as a major problem in the highland areas and major initiatives have been launched to tackle the issue. For example, the Ethiopian Highland Reclamation Study concluded that around 1900 million tonnes of soil are lost from the highlands each year, amounting to around 35t/ha/year (FAO, 1986). Similar pronouncements emerged from the early phases of the National Conservation Strategy process which emphasized the widespread nature of environmental degradation (Wood and Ståhl, 1989). The concern generated by such studies resulted in major campaigns from the mid-1980s to build soil bunds and terraces across the country, supported by massive food-for-work programmes (Hoben, 1995; Keeley and Scoones, 2000a). Similarly, soil-fertility decline has been highlighted as a significant constraint to agricultural production and food self-sufficiency (Wales and Le Breton, 1998), and major efforts have been made to encourage the wider use of inorganic fertilizers (Takele, 1996).

From the late 1960s, agricultural policy has been framed in terms of the need to 'modernize' Ethiopian peasant agriculture through a process of technology transfer. During the 1970s a series of integrated rural development programmes were established in different areas (Cohen, 1987; Ståhl, 1981). These later led to an agricultural extension approach based on a series of technology packages based on improved seeds and fertilizers. In recent years this technology transfer approach has been promoted first through the

Sasakawa-Global 2000 programme, and subsequently through a major extension push by the Ministry of Agriculture (Sasakawa-Global 2000, 1995).

This chapter reports on work (carried out in Wolayta in southern Ethiopia) which tries to set this debate within a particular local context. The chapter asks how farmers manage soils on their own farms, investigating what factors encourage soil improvement or result in soil decline, and exploring the links between local soil-management knowledge and practice and wider social, economic and political processes. In particular, the chapter asks how appropriate the policy approaches promoted over the past 30–40 years have been in improving rural livelihoods and ensuring sustainable management of natural resources.

Following an introduction to the study area, a brief environmental and agricultural history is offered which sets the contemporary debates in historical context. The way different events – both natural disasters, such as drought, and political and policy changes – have affected the way farmers have been able to manage soils is then examined through a series of multigenerational biographies. The next section turns to an assessment of land users' perspectives on soils and their management, presenting data on soil classifications and local understandings of soil-fertility change. The chapter then turns to the socio-economic dimensions of soil management with an examination of both survey data and case study material that show how different factors influence soil-management practice. The impact of such practices on soil-fertility is then explored with a presentation of data on nutrient budgets at farm and field level, as well as information on soil-amendment practices by different farmer groups. Finally, the chapter assesses the case study findings in relation to broader questions about the relationships between poverty, livelihoods and soil management with an examination of the different pathways of livelihood and natural resource sustainability. The institutional and policy requirements for ensuring more sustainable options for the future are then briefly reflected upon.

THE WOLAYTA CONTEXT: AN INTRODUCTION TO THE STUDY AREAS

Wolayta is a diverse area in southern Ethiopia (see Figure 2.1) characterized by dense human populations, particularly in the highland zones where, in some parts, population densities are as high as 500 people per km^2. Despite the apparent richness of the volcanic soils and the lushness of the vegetation and crops, the area suffers recurrent food shortages. For this reason the area has been labelled 'the land of the green famine' (Dessalegn, 1992). The vulnerability of the agricultural base makes investigation of soil-management questions central. Indeed, some commentators argue that it is environmental degradation that is at the root of many of the problems faced in the area.

Within Wolayta, rainfall varies from around 1200mm per annum in the highlands to around 750mm in the lowlands, although there are large interannual and interseasonal variations. The main growing seasons are between

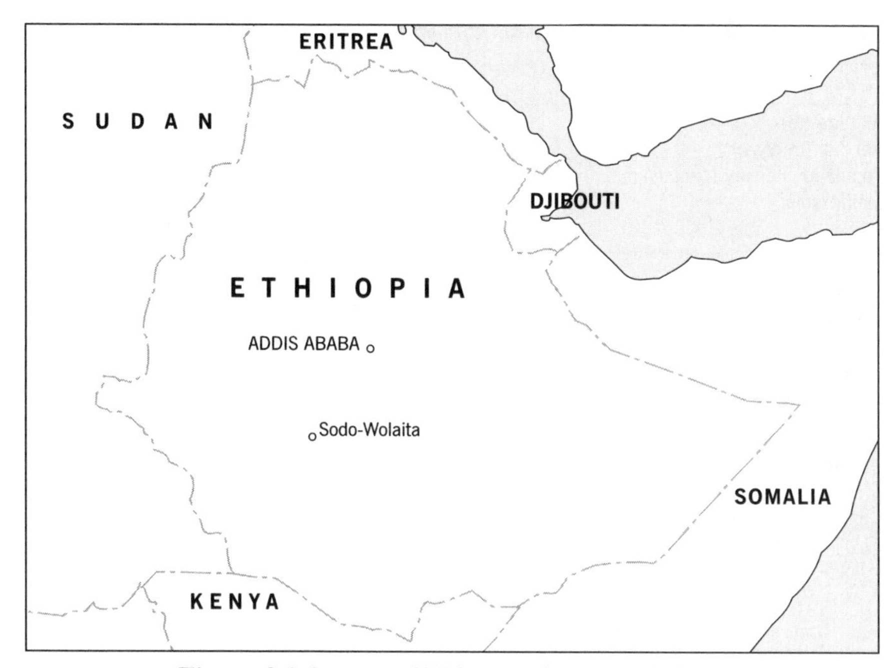

Figure 2.1 *Location of Wolayta study areas in Ethiopia*

March and April (*belg*) and between June and October (*meher*). Agro-ecological variation is accentuated by the undulating topography; there are large altitudinal variations within the study area, from highlands (over 2000masl (metres above sea level)) to lowlands (around 1000masl). Soils of the area are dominated by volcanic eutric nitosols, often with significant phosphorous deficiencies (Weigel, 1986; Belay, 1992; Gunten, 1993), although soils are highly variable on the ground (see below).

Livelihoods in the area are primarily agricultural, but significant additional incomes are derived from off-farm sources (Kindness, 1994). The agricultural system is characterized by a mixture of perennial, root, grain and tree crops, grown on small plots ranging in size from 0.1ha to over 2ha in the highland sites. Perennial crops include enset (*Enset ventricosum*) and coffee (*Caffa abysinica*); root crops include taro (*Colocasia esculenta*); grain crops are dominated by maize (*Zea mays*) and teff (*Eragrostis teff*). Trees, notably eucalyptus, are also grown on farm plots and are an important source of cash income. In the lowland settlement areas a similar farming system prevails, although there is a great emphasis on grain crops, notably maize, plus some cotton as a cash crop (see Kefala and Sandford, 1991). Moreover, the land sizes are considerably larger, with farm plots ranging from around 1ha to over 5ha. Livestock – cattle, equines and smallstock – are an essential component of both the highland and lowland farming systems. Oxen are critical for ploughing, while manure and urine from all animals are vital for soil-fertility management. In addition, the protein derived from livestock products is important in a high carbohydrate diet based on enset.

Table 2.1 *Profiles of the highland and lowland study areas*

Characteristic	*Highland*	*Lowland*
Altitude (m)	2100	1170
Rainfall (mm/year)	771–1743 (avg: 1272)	542–1636 (avg: 924)
Population density (people/km^2)	375	110
Major crops	Enset, maize, teff, sweet potato, taro	Maize, sorghum, sweet potato

This chapter reports on work largely carried out in two main sites in Kindo Koisha district, with one in the highlands (Gununo area) and one in the lowlands (Bele area). Table 2.1 highlights some of the major contrasts between the sites.

This work stemmed from a reflection on past studies as well as practical development experience in the area. The issue of soil-fertility had been raised by farmers regularly during a series of participatory appraisal exercises across Wolayta.[2] Ministry of Agriculture officials, NGO workers and others therefore became increasingly aware of the issue, but were at a loss as to how to react. Past attempts to intervene had seen little success. Subsidized credit for fertilizer, for instance, had been tried for almost a decade from 1971 under the WADU integrated rural development programme, but input levels dropped once the programme withdrew (see below). Similarly, attempts to encourage composting, agroforestry, alley cropping, green manuring and other types of organic-matter management had also seen limited success. A closer look at the dynamics of soil-fertility change, and the ecological, social and economic contexts of soils management was needed in order to understand the issue more clearly.

For this reason, a wide-ranging interdisciplinary study was launched during 1995, focussing on Kindo Koisha district. The study aimed to contrast highland and lowland settings, and take a differentiated socio-economic perspective to the analysis. Local understandings of soils were central, requiring a detailed, field-based approach to inquiry. A range of fieldwork approaches were adopted, which complemented reviews of existing data and archival searches. These included informal discussions with farmers, observation and participation in farming activities, a household-based sample survey of 100 farmers, resource-flow mapping at farm level, nutrient-balance studies on eight selected farms, and biographical and historical interviews with key informants.

AGRICULTURAL AND ENVIRONMENTAL HISTORY

In order to understand dynamics and change, it is important to set understandings of contemporary agricultural and environmental issues within a historical context. Contemporary images of Wolayta as overpopulated, famine stricken and environmentally degraded have not always existed.

In the early part of the 17th century, Manoel de Almeida observed how 'generally speaking this country is very fertile, for in some parts it yields two or three crops a year'. He notes 'not much woodland', but a range of crops, including wheat, barley, millet 'of many different kinds', sorghum, teff, enset and cotton, along with a wide range of fruit trees and medicinal and fragrant herbs. However, he goes on to note that despite such plenty, 'hunger is common' due to the depradations of pests, diseases, and, particularly, lawless armies. In addition, he notes that the lack of transport connections makes trade between districts limited (Beckingham and Huntingford, trans, 1954).

Until the last century, populations were low and land was abundant. Oral histories recall that houses were scattered across the landscape, with large areas of uncultivated land and forest in between. The plentiful wildlife provided an additional source of food for the local inhabitants. Ato Altaye from Fachana Peasant Association (PA) recounts:

> *It was my great-grandfather who came to this place from Kindo. At that time the place was all forest, inhabited by wild animals. The first settlers established their homes keeping some distance from each other. Each morning the people used to look out to see if smoke was still coming from their neighbour's house to ensure that they had not been attacked by wild animals during the night. My great-grandfather purchased this land from Takiso Amado, who was then the Wolayta king's son. At that time the dominant crops were sorghum, taro, enset and maize.* (Interview, 1996)

The area was part of the wider Wolayta kingdom, a centralized administration under a king; a highly hierarchical society characterized by caste and slave relations. During a series of confrontations through the 1880s and 1890s, King Tona resisted the attempts by the Amhara from the central highlands to take over this fertile and plentiful land. Vanderheym, a French commercial agent who accompanied Emperor Menelik in his final campaign against the Wolayta in 1894, was fascinated by the fertility and wealth of Wolayta at that time:

> *[Wolayta] is extremely fertile; numerous plantations of corn, wheat, barley, coffee, tobacco, cotton and millet surround the compounds of huts and make the country's appearance rich. Vegetation is abundant with figs, palms, spindle trees, sycamores, etc. Skins of animals, gazelles, lions or leopards, hung in the inside of the huts show how hunting was a major activity. They were a happy population, very self-sufficient, still living a biblical style of men to display their courage and heroism.* (Vanderheym, 1896 quoted in Chiatti, 1984, pp334–335)

Indeed, it was the agricultural riches of the Wolayta area which attracted the Emperor and his armies. The military campaigns of this period took their toll, as soldiers and followers were expected to live off local produce. Many also took advantage of their position, and looted considerable booty from the villagers. Vanderheym, for instance, reports that Menelik himself kept 18,000 cattle, plus a large number of slaves.

Other travellers in the region also confirmed the agricultural potential of the area. For example, Captain MS Wellby observed in 1901:

> *The soil, much of which is of red ochre colour, is extremely fertile, and those portions that are not covered with fine timber of crops, produce a multitude of flowers and good grass... The inhabitants of this district supplied me plentifully with sheep, eggs, fowl, ghi, bread, milk, teg and honey, and themselves appeared happy, contented and well fed. The district of Wolamo [Wolayta] has a fine climate, and should, with proper development and the facilities of communication and transport, become exceedingly prosperous.* (Wellby, 1901, pp140–141)

Stigand noted in 1910 that, although there were no inhabitants in the lowlands, the plateau areas were 'thickly inhabited', with 'a considerable amount of the country under cultivation' (Stigand, 1910, p294). He observed the typical Wolayta farming system with 'clustered plantations of the wild banana [enset]' around the huts.

Following the Amhara occupation of the area from 1894, there were, however, some major changes in land administration imposed which fundamentally affected land use and management. Four major types of land holding were designated under the feudal *gult* system of land rights. These were *lamia* (lands owned by farmers), *tafia* (abandoned land), *kaladia* (unoccupied land belonging to the state, and often allocated to service men, largely Amharas), and *gutara* (communal land for grazing or social events). The influx of people who opened up and occupied new lands began to transform the agricultural landscape. In addition, the severe taxation imposed by the new Amhara administrators put a major strain on agricultural production, pushing many to abandon their land and seek tenancy arrangements elsewhere under new landlords.

During a 1915 expedition by the British Consul, Arnold Hodson observed:

> *The districts of Kambata, Walamo and Baroda are extraordinarily fertile and productive, but enormous areas are left uncultivated on account of the exactions of the Abyssinians. A tax of one-tenth is levied on all produce, and this inevitably tends to prevent the people from growing more than is essential for themselves. With a just and stable administration, the total production of this part of Abyssinia could be increased many times over.* (Hodson, 1927, pp29–30)

The Italian occupation of the country offered a brief respite from these exactions, with attempts to establish a fairer taxation system to override the tribute labour arrangements and excessive grain demands that had emerged under the *neftegna-gebar* system. But this was short-lived, and the previous system quickly re-established itself with the return of the imperial regime under Haile Selassie in 1941.

However, despite these major political and administrative changes, the area was still seen by outside observers as 'rich and fertile'. For example, Scott, reporting on an expedition to the region in 1948–1949, noted:

> *The close cultivation, the hedges of eupohorbia round enclosures near the town, and the plantations of enset trees, with the hot bright sunshine glinting from their burnished leaves in the crystal clear atmosphere between the October thunderstorms, all contributed to an impression of richness and fertility.* (Scott, 1950, p128)

The agricultural success of the area was reinforced during the 1950s and into the 1960s by the boom in the coffee industry in the area. Soddo town was a central hub in the emergent transport networks of the south, making transport relatively easy and prices high. The market boomed, and many farmers profited greatly, investing in farm improvement as well as diversifying their income sources with the establishment of other trading activities. However, this boom was not to last. A combination of factors impinged – world prices declined, coffee-berry disease wiped out many plantations, and new transport routes opened up to the west, shifting the focus of the coffee trade.

By the late 1960s, doubts about the future viability of the agricultural system of Wolayta were beginning to be expressed. This was the turning point in commentaries on the area. No longer was Wolayta seen as a rich and fertile land, but one suffering from overpopulation, land degradation and poverty. Some of the earliest concerns were expressed by the governor of Wolayta at that time, Wolde Semayt. During the 1960s, Wolde Semayt had become convinced of the need for modernization of agriculture in the area. He argued that the old ways were inadequate and that major changes had to be encouraged through the transfer of modern agricultural technologies, combined with a process of resettlement to new lands which would be cleared in the lowlands.

This vision of transformation was to become the basis of the Wolamu Agricultural Development Unit's (WADU) activities for a decade from the early 1970s. Wolde Semayt was a key person in encouraging the World Bank to support the project. A series of appraisal reports were carried out in the early 1970s (WADU, 1976a, b, c) which attempted to diagnose the basic problems of the area. The reports concluded that production could only be increased if farmers shifted to growing more grain crops (notably maize and teff) with inorganic fertilizer. This would be supported by an ambitious programme of credit support. The agricultural improvement strategy would, in turn, be linked to a settlement programme which would encourage people to move to large plots in the lowlands. This would relieve population pressure in the highlands and allow new agricultural entrepreneurs in the lowlands to combine extensive grain cropping with cotton production.

In the early years, the WADU programme was implemented with vigour and commitment. The project director, Victor Burke, waxed lyrical about the potential of the WADU proposal of 1972, commenting that: 'The eyes of society have been opened to previously unimagined possibilities' and 'the complaisant peasant of two years ago [when initial demonstration work started in pilot areas] has been transformed' (WADU, 1972, pp29 and 50). The proposal identified fertilizer as the main intervention, noting that 'this is the most obvious need and most valued benefit' (WADU, 1972, p39), based on the assumption that 'it is generally not possible to grow an arable crop in

Wolamo without the addition of phosphatic fertilizer in some shape or form' (WADU, 1972, p16).

The crop surveys and project evaluations over the subsequent years show significant gains being made. Maize production expanded enormously in the early years, with hectarages increasing ten-fold in many areas (WADU, 1973). By the late 1970s, over 6000ha of land had been conserved with soil and water conservation measures, and nearly 6 billion Birr disbursed as credit for fertilizer purchase (WADU, 1978). Yields of maize had increased significantly on farmers' fields, with an average of around 3000kg/ha being realized in the 1976–1978 period (WADU, 1977). Many farmers recall this period positively. Reflecting on the reliable provision of services, one farmer from Molticho PA commented: 'WADU was like our father'. Boshe Chebo from Kindo Koisha recalled:

> *We had high crop yields with the application of fertilizer during WADU time. I have never seen such a high yield either before or after WADU. Our ancestors gave us the land, but application of fertilizer under WADU gave us the yield.* (Interview, 1997)

WADU was part of a wider agricultural policy approach hailed in the third five-year plan, launched in 1968. This identified the need for the commercialization and modernization of peasant agriculture through a process of technology transfer. A series of extension programmes complemented the integrated rural development projects such as WADU. From 1967 a comprehensive package programme was launched. This gave way in 1971 to a series of Minimum Package Programmes (MPP) (1971–1974, then 1974–1985) which promoted improved seeds and fertilizers through a credit-based programme. From 1985 the MPP was transformed into a wider programme of extension support (PADEP), although in this programme and its successor PADETES (Participatory Demonstration and Training Extension System), the emphasis remained on attempts to technically transform what was perceived as backward and inefficient peasant agriculture (Dejene and Mulat, 1995; Solomon, 1996). Over the period from the early 1970s some successes in this approach can be noted. For example, fertilizer use in Ethiopia has increased from a mere 947 mt (metric tonnes) in 1971 to 236275mt in 1996 (Mulat, 1996; Mulat et al, 1997; Worku, 1998). Equally, there has been an increased use of improved seeds, particularly of cereals, and some observed increases in yields.

In Wolayta, the WADU programme resulted in some major changes in agricultural strategy and, as a consequence, soil management. The focus on grain crops (and to some extent, cotton) resulted in a de-emphasis of the traditional enset and root cropping system, so often commented upon by travellers and other observers in the past. With access to relatively cheap fertilizer and new seeds, many farmers shifted attention from their garden areas and the building of fertility through manuring and organic-matter management, to investing in the monocropped outfields. Given the prevailing conditions, this was a highly sensible choice. But, in the longer term, as we shall see, this increased vulnerability, as the mainstays of the Wolayta agricultural system were undermined.

The undermining of the long-term strategies of soil management and improvement, linked to root crop production and enset gardening, became even more apparent during the late 1970s and into the 1980s, following the Ethiopian revolution of 1974. The most dramatic consequence of the revolution was the land reform proclamation of March 1975 which attempted to abolish the landlord–tenant system and redistribute land to all peasants on an equitable basis. For former landlords this was a hard time. A farmer in Molticho PA, now aged 70, commented:

> *The revolution came and things changed. Land was distributed and we all had to work hard for some years. I did not join the cooperative so they considered me 'a feudal' and took my land. At that time I had a corrugated roofed house with three rooms. I was seen as a rich person. So I was forced to reduce the size of my house, and only farm a small portion.* (Interview, 1997)

Even though many of the ideals of the land reform were never realized, because local negotiations among particular former landlord farmers and peasant association leaders always produced compromises, the reform did provide land to a substantial group of people for the first time. This opened up new opportunities for those who were previously tenants to invest in their own land. However, at the same time, the reform, and the subsequent policies of villagization instituted in the 1980s, introduced a new sense of uncertainty. Since this time, farmers have never really known whether the state will require the land to be redistributed again, or whether the authorities will require people to move to establish new villages. For a long period, this sense of insecurity has discouraged many from investing for the long term in such assets as soil-fertility or on-farm trees. One farmer commented:

> *Our worry at the moment is insecure land ownership. We are always in fear of redistribution. I don't like to see a rope lying on my land. Land is our life and one should not expect a farmer to sell his land and become a labourer. If we sell our land what are we going to leave to our children?* (Interview, Mundena PA, 1997)

Villagization, in particular, was a severe blow to many. In Wolayta, the villagization policy was carried out largely in the lowlands where planned consolidation or linear development of settlements was more feasible than in the crowded highland areas. This meant that people were forced to abandon their lands which they had been investing in since they had arrived as part of the settlement programme over a decade before – enset plantations, garden areas and the rich soils associated with homestead sites were left behind in areas now allocated to other farmers' fields or to grazing lands.

The latter part of the 1980s was a low point in Wolayta's history. The subsidies of the WADU project had long disappeared, and government service provision was no match; land tenure insecurity continued to create uncertainty; and the escalation of the war in the north meant that enlistment to Mengistu's army became compulsory for all young men. This was a time of fear, uncertainty and increasing poverty. It should not have been a surprise that the failure of the rains in 1984 resulted in major famine. The 1984–1985 famine in Ethiopia gained worldwide attention, although most of the focus was on the north. Wolayta suffered badly too, and huge amounts of food relief had to be distributed. Even so, many people died. Those who survived had often disposed of most of their assets – enset plantations, livestock, jewellery, land – leaving them little to fall back on the next time a difficult time arose (Pankhurst, 1993). Since then, such times have come more frequently as people have become more vulnerable, with serious food deficits being recorded in the area repeatedly through the late 1980s and 1990s. In most years some type of food relief or food-for-work campaign has been organized by government, NGOs and others (Jenden, 1994; Webb and von Braun, 1994).

The expansion in food relief during the 1980s was associated with a growing concern with environmental degradation and, in particular, soil erosion. With the production of the Ethiopian Highland Reclamation Study (EHRS) in 1986 (FAO, 1986) a justification for substantial investments in soil erosion control was made, with dramatic claims about soil loss being offered.[3] The Soil Conservation Research Project (SCRP) established in 1981 with Swiss government support and under the leadership of Hans Hurni was particularly influential in presenting the scientific case for increased attention to soil erosion issues (Hurni, 1994). The research site at Gununo in Wolayta was one of a network of initially eight stations which were established to monitor soil loss and the impacts of soil erosion control measures.

The argument for urgent action on soil erosion linked well with the need to channel food aid through major public works programmes. During the 1980s, soil conservation works became the hallmark of such programmes, most notably through the collaboration between the Ministry of Agriculture and Natural Resources and the World Food Programme's 2488 project. While much of the focus of the EHRS and the subsequent food-for-work campaigns was on the northern highlands, the south was also seen to be in need of such support. The promotion of physical soil conservation measures – particularly large, fixed and regularly spaced bunds – was a major feature of extension advice through the 1980s in Wolayta and beyond.

With the overthrow of the Mengistu regime and the installation of the EPRDF government in 1991, some major policy changes were announced that reinforced initiatives made by the previous regime since 1988. These signalled the opening up of the economy and the encouragement of a liberalized agricultural sector as part of a broader structural adjustment of the economy. An agricultural-development-led industrialization strategy was announced which again put a strong emphasis on technology transfer and agricultural modernization. With the devaluation of the birr in 1992, a series of liberalization measures followed, including the encouragement of reforms of the fertilizer marketing system. The Agricultural Input-Supply Enterprise replaced the state-run AISCO which had been the sole source of fertilizer supply in the country. The encouragement of private entrants into the fertilizer market was also aimed at encouraging competition and reducing costs to farmers. However, the reduction of subsidies and their final elimination in 1997 has had the effect of increasing fertilizer costs, as the marketing system remains weak and inefficient (Mulat et al, 1996, 1997).

From 1993 a major influence has been the Sasakawa-Global 2000 programme which has promoted a credit-supported package of seeds and fertilizers. This strategy of 'aggressive technology transfer' (Borlaug and Dowdswell, 1994) was taken up by the Ministry of Agriculture as part of the new extension strategy (PADETES) which was aimed at linking a training, visit and demonstration plot-based extension system with the SG–2000 fertilizer and seed credit package. This approach has had a wide reach, particularly in the higher potential zones of the country. In the Wolayta study areas an increasing number of farmers are making use of the credit and input supply offered, in part because there are few alternatives. Although the successes of the exceptional rainfall year of 1995 have been widely trumpeted, the approach does not suit everyone every year, as one 65-year-old farmer from Doge Shakiso Peasant Association commented in 1997:

> *At the beginning those who used the extension package produced a lot. I was attracted so I volunteered last year. However I found it was not worth it if the rainfall was not conducive. Even if you produce more than without the package, the extra income goes back to repaying the debt. If you don't pay you can be sent to prison or the whole PA will not get fertilizer. This year the PA forced us to receive the extension package, and they told us there would be no other credit. I don't have cash to buy fertilizer from the free market, thus only for the sake of credit I had to take from the Global extension package.* (Interview, Doge Shakiso PA, 1997)

The lack of alternative credit options remains a major problem for farmers in the study area. The collapse of the service cooperatives as credit providers, the limitations of the Ministry of Agriculture system, and the absence of rural banking options, make the extension package perhaps the only option.

The security of land tenure has been a recurrent theme (see Dessalegn, 1994a) and remains a major issue in Wolayta. Although no fundamental changes in tenure law have been put forward, confidence in land security has risen at the local level, especially following the reform of the peasant association and service cooperative system. However, doubts remain particularly over informal 'contracting' arrangements for loaning land to others. One farmer commented:

> *I am more worried at the moment... We have no clear idea about our land ownership. During previous regimes, even the Derg regime, there was no fear to contract, though informally, a field for more than three years. At present we are in fear of losing our land.* (Interview, Kindo Koisha, 1997)

Thus changes in political regime, in land tenure arrangements, in marketing opportunities, and in extension policy have all combined to change the contexts for farming and livelihoods in Wolayta over the last century. The way such broader contextual changes have affected the way farmers have managed their land and their soils in particular will be explored in the following section.

BIOGRAPHIES OF CHANGE: THE INTERTWINING OF SOCIAL, ECONOMIC, POLITICAL AND BIOPHYSICAL PROCESSES

This section examines how, over several generations, changes in soil management have been influenced by the interaction of the sort of broader contextual factors discussed above and changing social and economic conditions faced by individual farm families. Three biographies are presented which trace the changing fortunes of different families, now classified as poor, rich-medium and rich, over several generations.

Case 1: Oda Orchile Sigo, poor, age 40

My grandfather had a lot of land and livestock and lived in this area before the arrival of Menelik and the Amhara. Some time later, a feudal lord from Soddo Zuria *woreda* confiscated a large portion of my grandfather's land, leaving him with only a small plot. However, the soil was fertile and he could do some cultivation there. Also the feudal lord took all his other assets, leaving him very poor. The lord also took my father, Orchile, into servitude in Soddo, where he remained for seven years until the Italian invasion. The Italians freed my father and he returned to his people and the remaining land. By that time, my grandfather, Sigo, had already died, and my father was able to inherit the small, but fertile, plot of land. Nevertheless, my father's brother, Kubanche Sigo, chased my father, who was young, poor and had not lived in the area for some time, to another area which was infertile. I, therefore, inherited infertile farm land and few assets following the death of my father. I was a victim of my family's history of being dispossessed by those in power. However, during the land reform I did manage to get some more land. At this time I started to transform my infertile land, but labour shortage, oxen shortage and poor health prevented me from achieving this. The last 20 years have therefore been very bad for me.

Case 2: Ato Geta Nasa Bala, rich-medium farmer, age 45

My grandfather lived in this area long ago. He was a rich man, owning many cattle and much fertile land. He was a warrior and a local chief. He had power to demand labour and manure as tribute from the people living around him. When he demanded manure, people would pile it up in their own homes and then carry it to his field. He even used to apply manure to his teff field. Because teff does not do well with much manure at one spot he demanded that the people come and place it in small amounts across the field. As a result his whole field was very fertile and he reaped a lot every year.

My grandfather, however, was not well liked, as he was a harsh man. He was later killed by his relatives who took his land on his death. My father nevertheless inherited some good land, as well as some livestock. He worked hard and built up his wealth, and became a rich land owner during Haile Selassie's time. In addition to being a farmer, he was also a trader, acting as a local butcher. Hence, through earning more income he was able to buy cattle and additional land. He had three wives, six sons and four daughters. In order to increase his labour force for his growing farm area, he fostered four young boys, who provided labour until they established their own households after his death. Sometimes, when labour demand was exceptionally high he used to call work parties (*dagwa*, *zayia* and *urpia*) to his land. For instance, work parties were often called when hoeing the field during the dry season, around October and November, in order to break the soil, and allow the sun's heat to increase the fertility. Despite having plenty of labour, as well as manure, he was unable to cultivate all the land he owned. So he allocated some of his fields for share cropping where others planted and tended the land, and paid him a share.

By the time of his death, he owned a large portion of land and many cattle. He was able to distribute adequate land to all his six sons, including myself, as well as a number of cattle to each of his children, including his daughters. At that time his daughters, my sisters, were also entitled to inherit land and livestock. But at the time of the inheritance negotiations the revolution came. This changed everything. My sisters no longer could claim land, but myself and my brothers got the land. A few years later the land reform policy was put into practice. This meant that my land was reduced a lot. This was a terrible blow. However, by my own effort I managed to acquire some more land through purchase. I purchased both the area I lost through land reform and some more in addition. At that time, I was gaining good money through coffee trading in the area. I bought coffee at one time from the farmers, stored it and later sold it at a good price. This money, together with crop sales, allowed me to buy more livestock and more land.

I now have three wives and ten children, but I still suffer from labour shortage. Unlike in the past, *dagua* is no longer viable because we plough the land with oxen, and people just demand too much food and drink. However, I do sometimes host *zayia* for hoeing some plots. I would have been happy to foster children for the sake of labour like my father, but the Derg regime disapproved of this, and the tradition has reduced up till today. Despite these problems, I have managed to improve the soil in my land. In some parts, it is even more fertile than when my father was farming. Some of the lands which I bought, however, were infertile (*lada*), but I am working hard to improve it. But the main problem I am facing now is that my livestock have been struck by disease in recent years. This means that it is difficult to plough all the land and the amount of manure has decreased. During the last three or four years, therefore, I have seen my wealth declining. In contrast to the days when I had money from coffee, large numbers of cattle and high yields of crops, I now am only self-sufficient.

Case 3: Ato Altaye Anaro Balango, rich, age 80–90

My grandfather came from Kindo at a time when land was plenty and the forests were inhabited by wild animals. But conditions changed drastically after the coming of the Amharas. Through taxation and other tribute, Amharas brought great hardship for the people of Wolayta. Some feudal lords were particularly bad and we had to abandon our lands and move to another area looking for relatively better masters. I abandoned the whole of my land here in Fachana and most of my land in the highlands, so that I did not have to pay so much tribute. The land was then taken over by other tenants chosen by the feudal lord. However, during the Italian occupation, I reclaimed my lost land here. But I did not return to this area until later, and left the land to be managed by tenants.

With the arrival of the Derg regime land reform was declared. Through this I lost most of my land in the highlands. But here in Fachana the PA leaders were sympathetic and I did not suffer much. I returned to my land here at this time. I tried to buy the land back in the highland, but the people who were allocated it sold the land to someone else. I was annoyed by this, and took the case to the *woreda* administration, but the case is still not solved.

WADU brought the idea of service cooperatives a few years before the revolution. It also brought fertilizer as a result of which yield increased. The service cooperatives brought some good ideas, but they were also corrupt. When the Derg regime collapsed, the officials stole money and built their own private wealth. The current government failed to investigate the situation. Now we cannot afford fertilizer and it is not even worth thinking about fertilizer, especially because we don't have labour or oxen to manage the land well.

Emerging themes

A number of important themes emerge from such accounts. Soil management can be seen to be intimately intertwined with the waxing and waning fortunes of individuals and families. In 1997, when these interviews took place, these individuals may have been classified as relatively richer or poorer and, because of available assets of land, labour, oxen, credit and so on, differently able to manage their soil-fertility. But such conditions may not have pertained over the longer period. As the case studies illustrate, some individuals may be better or worse off than their fathers or grandfathers, or compared to earlier in their lives, due to a range of different factors, many out of their control. The cases highlight a number of the key events mentioned in the historical review as particularly important. Changing land access and ownership rights is a recurrent theme. Whether feudal expropriation, land reform or changing inheritance rights, particular events can be seen to have had a major impact on the potential for soil management. Such externally-imposed changes, the result of shifts in the broader political order in Wolayta, thus combine with more local processes and the changes in the asset base of individual households.

As the biographical accounts recall, attempts to build up the necessary labour or oxen resources to effectively manage and indeed improve the soil are affected by a variety of factors. For example the expansion of tsetse impact has had a major impact on the ability of farmers to accumulate livestock, especially in the lowlands. The series of rainfall failures in the 1980s and 1990s, as we have already noted, had major impacts on people's asset base, as key resources had to be run down in order to survive. Changing institutional arrangements – for example the observed decline in work parties for communal labour – also have an impact on soil management. External interventions may also be important. Clearly, the WADU period is the most significant, as the subsidized credit arrangements made the purchase of fertilizer possible for a greater number of people than ever before and so changed the priorities in soil management and agriculture for a period.

Thus, over time, it is the conjuncture of exogenous events – both deriving from natural and political processes – with local chance happenings – an illness, a death in the family, a dispute among relatives – combined with ongoing processes at household and community levels – of asset accumulation, demographic change and institutional transformation – that affect the ability of individuals and groups to manage soils in certain ways at any particular time. Thus the changing soil-fertility of a plot of land may be as much to do with political, social and economic change as with biophysical parameters. Retracing the event history of a particular plot of land thus requires linking changes in biophysical properties – for example the expansion and contraction of manured garden areas over time – with the changing fortunes of the individuals and households who manage the land. Such histories show that there is no simple pattern of inexorable decline or constant improvement in soil properties, but combinations of events, the result of a particular constellation of circumstances, may result in relatively rapid transitions between soil states (see Chapter 1).

LAND USERS' PERSPECTIVES ON SOILS AND THEIR MANAGEMENT

How is the changing soil environment viewed, and, in turn, managed, by local land users? This section investigates this question by looking at farmers' classification of soils, their conceptualization of soil-fertility change and their daily management of soils. A series of mapping exercises with farmers in highland, mid-altitude and lowland sites highlighted the classification of soils in the area (Figure 2.2).

A total of 10 different soil types were identified by farmers, differentiated by a range of criteria, including colour, workability, erosion susceptibility, moisture conditions and fertility status. These soil types were subsequently analysed in a laboratory which considered particle-size distribution, organic carbon, total nitrogen, pH and colour. A summary of these results is contained in Table 2.2.

Although a variety of different soil types are recognized, these are not static. Soils can change and be transformed over time, farmers argue. Two concepts are key: *arada*, which denotes fertility and *lada*, which denotes infertility (Table 2.3). In the comparisons farmers make, soils are described as rich

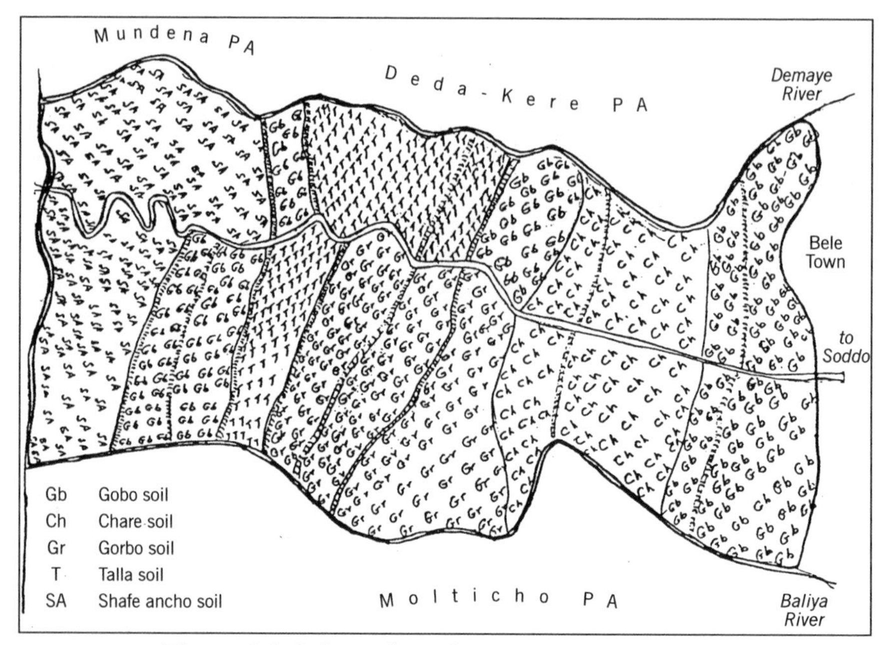

Figure 2.2 *Soil map drawn by farmers, Fagena Mata PA*

or poor, fat or thin, strong or weak. Soils are regarded as living and created, and often parallels are made with the human or animal worlds in describing their status (Data, 1996; Data and Scoones, forthcoming). Soils of different fertility are also associated with different landscape positions (gentle or steep parts of the slope) or niches within the farm system (near the home or far from it). Equally, fertility levels are associated with the type of management and form of cultivation, with fertile soils invariably being well-manured soils created in gardens through intensive hoe cultivation, mounding and active management and incorporation.

Change between these states may occur within any of the 10 soil types identified, no matter what the original state was. For example, some soils may be infertile when the land is cleared, but be transformed into productive land through cultivation. This was the case on Ato Tomma's land. He explains:

> *When I inherited this land, the soil-fertility was poor (*lada*). The garden (*darkoa*) area was very small. The soils in almost all the other plots were dying and very 'skinny' (*gilka*). Luckily I started working on them before they died (*haiqoa*) and were completely ruined. Now I have expanded my garden (*darkoa*) and also fattened (*anqara*) the soils in my land through my own efforts. Currently the soils in my land are more fertile than most of the farmers around here. This was made possible through continuous investment and work (*ossua*) and the effective keeping (*nagua*) of the land. Soils need continuous tending and care just like one does with children or cattle.* (Interview, Lasho PA, August 1995)

Table 2.2 *Local classification of soil types*

Local name	*Local description*	*Laboratory analysis*
Bossolo	Naturally poor; easily washed away; grassland areas.	High sand content, low organic carbon, pH of 5.5.
Kareta	Black soil; high moisture retention; prone to waterlogging; hard to work.	Deep soil, pH of 7.4, loam texture on top (26 per cent clay), more clay lower (71 per cent below 83cm). Fair C:N ration of 12.6 in top soil.
Talla	Dark red to black soil; water stress problems; difficult to work: cracks when dry, sticky when wet.	Vertisol with high clay content (44 per cent), organic carbon around 2 per cent, high C:N ratio (up to 20 in top 50cm), pH 6–6.6.
Bokinta	Shallow, found on steep slopes; poor fertility.	Limestone derived, clay content 54 per cent, C:N ratio of 13 in top layer, pH 6.3 in topsoil, but lower at depth.
Gobo	Reddish; easy to work; relatively fertile; deep soil; good for cropping, although some problems of water stress.	High clay content in lower horizons (rising from 30 per cent to over 60 per cent), pH 5–6.7, organic carbon around 2.5 per cent, variable C:N ratio (15–20 in top 50cm).
Salisatya	Poor soil; infertile; found on sloping land; rehabilitation impossible.	Shallow, pH of 4.8, high sand content (85 per cent), low organic carbon (1 per cent).
Barta	Light soil; often quite infertile, but responds to fertility inputs; high moisture retention; good for crops and trees.	High clay content (49 per cent), high organic carbon (3.3 per cent), C:N ratio of 26, pH 5.1.
Gorbo	Good agricultural soil; found on flatter lands; good water retention; easy to work.	Thick A horizon, 68 per cent sand, higher clay content at lower levels (up to 66 per cent), organic carbon content around 2.5 per cent.
Shafe ancho	River gravel or sand; poor water retention; crops perform badly.	Thin soil, pH of 7.5, clay content 16 per cent, organic carbon of 1 per cent.
Charya	Swampy land; problems of waterlogging.	Gleysol type, sandy loam, clay content of 19 per cent, organic carbon around 2.5 per cent.

Source: Ejigu et al, 1997

Similarly, soils may deteriorate through neglect and poor management. In the highlands, erosion may be a problem if bunds are not built. The topsoil quickly washes away with the heavy rains, leaving poorer subsoil.

A range of indicators of soil-fertility status exist which farmers use to judge the soil's requirements. Soil colour and its thickness are important, with deep, darker, black or red soils being known as fertile. Weed flora are also useful indicators of soil-fertility. Areas with a high prevalence of *Urochloa panicoides* (*maga matta*) or *Commelina bengalensis* (*dalasha*), for instance, are recognized as possessing higher fertility than areas with *Galinsoga parviflora* (*bisida*), *Phalaris eragrostis* (*girolia*) or *Digitaria* spp. (*petta*, *lichea*). Ultimately crop performance is used to judge whether certain inputs are needed. If crops are stunted,

Table 2.3 *Farmers' perceptions of soil-fertility: some contrasts between fertile and infertile soils*

Arada (fertile soils)	*Lada (infertile soils)*
On gentle slope	On steeper slope
Deep top soil	Shallow top soil
Black	Red
Rich	Poor
Fat	Thin
Strong	Weak
Manured	Not manured
Near house	Far from house
Hoed	Ploughed

Source: Data, 1996

have yellowish leaves and never yield well, even in a good rainfall year, farmers say that the soil must require attention.

A variety of ways of improving soil-fertility are used locally. The preferred route is the application of manures and composts, derived from livestock wastes, household refuse, crop residues and collected biomass, including leaf litter. This, farmers argue, improves yields over a number of years, and, in the longer term, results in a sustained improvement in the soil's structure. Livestock manure is particularly highly regarded. The dung and urine, combined with hay and crop residues, is regularly collected and piled in the garden area. Livestock are kept within the Wolayta house and are regarded almost as part of the family, seen as especially valuable because it is their manure, particularly that from cattle, that allows agriculture to thrive and people to eat. One farmer commented: 'manure is so valuable it should be put in a store like grain'; others joked that divorces had been known to occur over how manure was handled.

Other forms of organic matter are seen to complement livestock manures, but they cannot replace it. Those without cattle are in a particularly hard position. While they may make use of a variety of trees for leaf litter, and assiduously recycle crop residues, sometimes through a compost pit, they can never really compete with those who have cattle, sheep or goats. For this reason, all such farmers strive to get access to livestock, either through purchase or through share-rearing arrangements.[4]

Inorganic fertilizer is well known in the area, as the WADU project promoted the use of DAP (di-ammonium phosphate) extensively throughout the 1970s. Farmers recall this era with some fondness. Fertilizers, they argue, solved their problems – fertilizer was cheap and they got good yields on their maize and teff. However, since the WADU period, fertilizer use has dropped off. Official recommendations continue to suggest an application of 100kg of urea and DAP per hectare, and this is the basis of the now widely promoted Sasakawa-Global 2000 package. However, other research carried out from the late 1980s has suggested a more fine-tuned set of recommendations based on variations in soil type, moisture availability and crop requirements (ADD/NFIU, 1988, 1992).

But agronomic questions are only part of the equation. Due to increases in fertilizer costs as a result of the removal of subsidies and the parallel decline in output prices due to increases in supply, a sharp decline in the profitability of fertilizer use has been noted during the 1990s (Mulat et al, 1997). The value–cost ratio declining to below two (the usually accepted cut-off point for widespread adoption) following the removal of subsidies from 1996. While in terms of simple marginal returns fertilizer use remains profitable for most crops, even if optimal application rates are lower than recommendations (Kindness and Sandford, 1996; Worku, 2000), the institutional and organizational problems of effective and timely input supply and the availability of credit remain major issues constraining more widespread use.

Most studies point to a range of factors which influence fertilizer adoption. National surveys, for instance, point to literacy, access to all-weather roads, access to banks, extension service coverage and availability of labour as critical factors (Croppenstedt and Mulat, 1997; Croppenstedt et al, 1998; Akilu, 1980). In the Wolayta context these factors are clearly important, with significant constraints in relation to some or all of these factors being evident for a large majority of households in the study sites. The need to improve rural credit facilities, reduce the costs of supply through marketing reforms and infrastructure improvements, and encourage a greater labour market (Mulat et al, 1996, 1997) are all as relevant to the Wolayta setting as they are more broadly.

However, in the wider policy literature there remains an assumption that an increase in inorganic fertilizer application rates is the sole route to agricultural improvement. How do local people see fertilizers? While experiences from the WADU period are clearly positive, there are certainly reservations expressed about the similar SG-2000 package programme. In local conceptions, fertilizers play a particular role in soil-fertility management, and by no means the most important one. In Wolaytinia fertilizers are termed *talia*. This word has a number of connotations, ranging from poison to medicine. The term describes how fertilizers can provide a quick and effective response. However, the effect is temporary and limited. Like medicine, farmers argue, it cannot be taken over a long time, otherwise ill effects, including 'addiction', may be seen. While farmers would not dispute the effectiveness of fertilizers, they cannot be compared to manure, they argue; only through manuring can soils really be improved in the long term.

In the highlands and the mid-altitude zones where fields are found on sloping land, measures to prevent soil erosion must complement such amendments. Terracing and bunding have long been part of farming practice in the area. For example, Stigand observed mountain sides where 'terraces are built up with a stone wall at the lower end to prevent the soil being washed off the surface' (1910, p292). Such indigenous soil and water conservation structures are flexibly designed, making use of available materials. They were often shifted around a farm, focusing on problems where they arose. Fertile soil which built up behind a bund could then be spread out when the bund was moved to another spot, creating a highly productive niche in the field. But formal research and extension has instead encouraged more permanent measures, with large bunds being dug across the contour. In one part of the study area, a large SCRP-supported research project was initiated which treated one part of the catchment with a standard form of bund to test their effectiveness in reducing soil erosion. While farmers confirm that such measures are indeed effective at reducing erosion, they highlight other problems with such an approach. These include the amount of land that is taken up with their construction, the fact that such permanent structures become a favoured habitat for snakes, mole rats and other pests, and the problem of good soil accumulating behind the bund but not being used.[5] In recent years, a less coercive approach has been adopted by government agencies with respect to soil conservation. As a result, farmers have begun to adapt the bunds built on their land through research or development projects, making them more like the terraces and bunds of the past.

But in order to understand soil-fertility management in the Wolayta context, a differentiated view of the farm is essential, as different parts of farmers' fields are treated in different ways. Understanding the spatial patterning of soil management is therefore key. A number of different sub-components are evident in virtually all Wolayta farms in the highlands. These include the house site (*katesa*) and front yard (*karea*) which are surrounded by small spice gardens (*darincha*) and a larger enset garden. Further away from the house is found the maize garden (*darkoa*), which often has a mounded area planted to taro at its furthest edge. Beyond the darkoa garden is the outfield (*shoka*), and, at the furthest extreme, the grass and tree patch (*mita gidiya*). Figure 2.3 shows a farm map drawn by a group of highland farmers, which highlights each of these different components and some of the major flows of resources affecting soil-fertility between them.

A spatially-differentiated look at fertility management shows how manure is concentrated in two sites, the enset and *darkoa* garden, while fertilizer is largely used in the outfield (*shoka*). This has important implications for the way soil-fertility transformation at a farm level is viewed by farmers. The garden areas (enset and maize *darkoa*) are intensively managed, often with hoe cultivation. They receive often quite large amounts of manure and other biomass and, as a result of years of such investment, highly fertile, rich soils with good structure and high productivity may be created. The farmers' aim is always to expand this area outwards, pushing the frontier of highly fertile soil out into the poorer *shoka* outfield area. This process takes many years, and involves a sequence of investments. This starts with the planting of taro root on the margins of the *shoka* areas. Mounds are created and much biomass from previous maize, sorghum or teff residues is incorporated. Manure is also added and a fertile band is created. The mounding and associated hoeing, farmers argue, encourages soil improvement, and the taro, with its high biomass residues, is, they say, the ideal crop to encourage the transition from *shoka* to *darkoa* (Eyasu and Scoones, 1999).

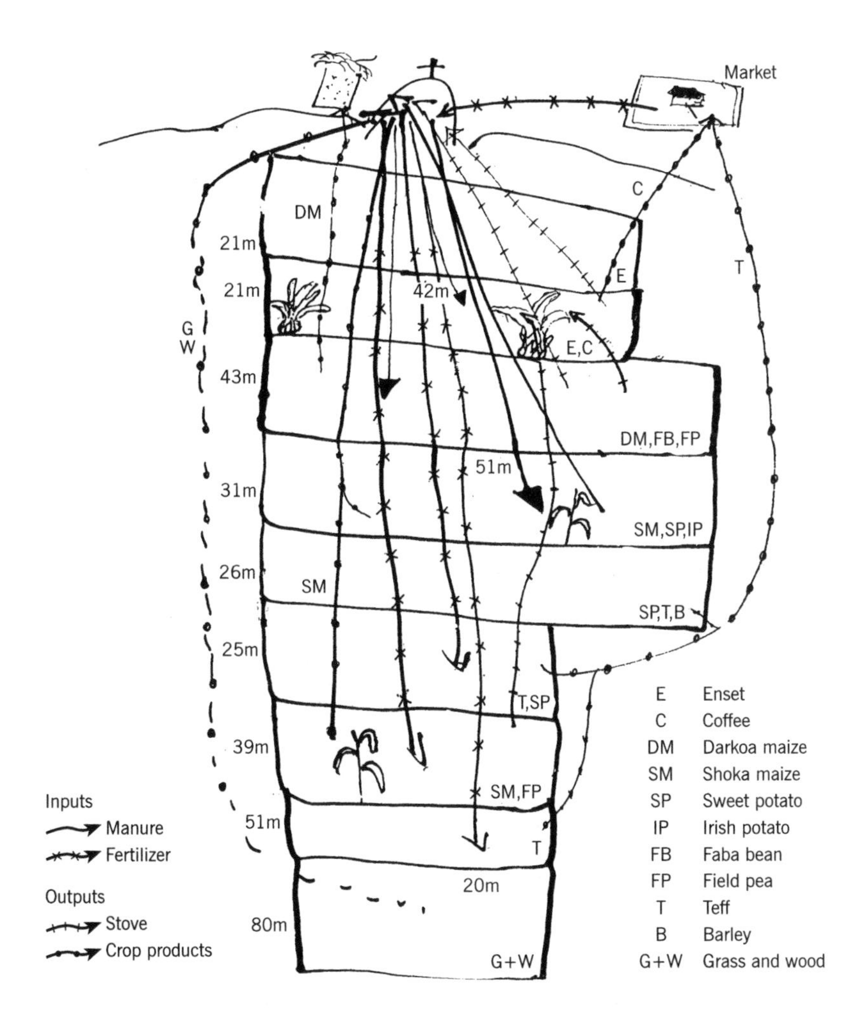

Figure 2.3 *Resource-flow diagram*

SOCIAL AND ECONOMIC DIMENSIONS OF SOIL MANAGEMENT

The ability of any farmer to transform the land in this way depends on a variety of factors. Previous sections have illustrated how soils are conditioned by socio-economic contexts – the continuous product of the interaction of the ecological and the social in the everyday practice of farming. This section dissects this interaction through a look at who does what in relation to soil-fertility management in the case study area.

As we have seen, the household's asset base is a significant factor in influencing soil-fertility management approaches. As discussed above, individuals' asset portfolios change over time, but some broader trends can be observed over time in the study areas. Table 2.4 contrasts the overall average situation in

Table 2.4 *Changes in household capital assets, 1971–1998*

	Highlands (Bolosso)		*Lowlands (Bele)*	
	1971	*1998*	*1971*	*1998*
Land holding (ha)	1.2	0.6	5	3
Cattle ownership	4.1	3.2	7.4	2.9
Oxen ownership	0.9	1.3	2.9	1.2
Family size	–	5.9	4.8	5.7
Church	Orthodox – 54%	30%	15%	11%
membership	Protestant – 17%	63%	31%	77%
Basic literacy rates	5.4%	35%	4%	36%

Sources: 1971 baseline from WADU surveys; 1998 data from IDS study (Carswell et al, 2000). The highland site is Bolosso nearby Gununo, and the lowland site is Bele/Mundena

the early 1970s with that pertaining today in relation to a number of key features.

A number of important trends are observed. Overall, total land holdings have declined. Increases in the rural population has meant that subdivision of land holdings has continued with the result that in the highlands the average land holding is now only 0.6ha, with 43 per cent in Bolosso having less than 0.5ha and 84 per cent having below 1ha of land in 1998 (Carswell et al, 2000), although today there is a more even distribution of small parcels of land compared to the 1970s, when landlord–tenancy relations meant a highly skewed distribution. In the lowlands, the original allocations of land under the settlement scheme have been subdivided through inheritance, although the actual amount which is cultivated has remained more or less stable through the period (around 2.5ha). Despite land reform, however, a range of sharing and contracting relationships exist in respect of land. Thus larger land owners can contract out excess land or enter into sharecropping arrangements with those who lack sufficient land (Carswell et al, 2000).

Although the total density of livestock in both sites has increased over the period (especially in the highlands), the number owned per household has declined. This has been particularly so in the lowlands where the impacts of trypanosomiasis have been especially felt since the 1980s.[6] With draft power being a key constraint, the herd composition has shifted more towards oxen than in the past. As with land ownership, there appears to be a more even distribution of cattle ownership today than under the feudal arrangements of the 1970s. In 1971, for instance, tenants owned on average only 1.75 head, while the overall average was 4.1 animals. Despite such changes, 19 per cent of households in the highlands and 9 per cent in the lowlands are recorded as owning no cattle at all today (Carswell et al, 2000). Thus the redistribution of livestock within and between sites through loaning and sharing arrangements remains an important feature, with 68 per cent of households in Bolosso being involved in either *hara* or *kotta* arrangements in 1998 (Carswell et al, 2000). In the lowland areas more cattle are recorded as being held rather than being owned, while the opposite holds in the highlands. This suggests a net flow of livestock from the highland areas to the lowlands although, due to risks of trypanosomiasis in the lowlands, most people comment that this practice has declined over time.

In terms of the labour and skills associated with human capital assets, the amount of labour available at a household level has remained similar over time, although data on household membership is difficult to interpret, with migrants, fostered children and labourers sometimes being included as members and sometimes not. Qualitative information suggests that labour remains a key constraint and, given the highly differentiated division of agricultural tasks, this has an important gender (and to some extent age) dimension. One of the major changes that has occurred is the increase in basic literacy in both sites. Other surveys have observed correlations between literacy levels and the adoption of a range of agricultural techniques promoted by extension, most notably fertilizer use (Croppenstedt et al, 1998; Teressa, 1997), suggesting that the observed increases in literacy rates may have an impact on soil-fertility management choices.

Changes in institutional contexts are observed across time. The changing religious affiliations of residents of the case study areas is a notable feature, with a major increase in membership of new Protestant churches being observed. This has affected a range of social networks and institutional arrangements in local society. For example, traditional work parties and associated social networks have shifted towards collective activities associated with church groups (Data, 1996). While membership of *iddir* (Dejene, 1993) has remained high in both sites (over 80 per cent of households having some association with such a group both in 1971 and 1998), other group-based institutions, notably dago work parties, have declined from over 80 per cent of households being regular participants in both the highlands and lowlands in 1971, to 48 per cent and 28 per cent in the highlands and lowlands respectively in 1998 (Carswell et al, 2000).

So how does changing access to such livelihood assets affect soil management? Over the period from 1971 to 1998 a number of major changes in soil-fertility management practice have been observed. In quantitative terms, fertilizer use has risen from effectively zero in 1971 to 94 per cent of household users in the highlands and 86 per cent in the lowlands in 1998 (Carswell et al, 2000). At the same time manure applications have reduced, reflecting declining access to cattle and manure. These have become more concentrated on enset, taro and sweet potato gardens in the highlands, with maize in the outfields losing out.

Table 2.5 highlights a range of socio-economic characteristics of richer and poorer groups of farmers in the highlands and lowlands derived from a survey carried out in the study areas in 1996.[7] Clear contrasts exist in a range of key factors of production – labour, land, livestock and cash income – across agro-ecological zones and across wealth groups. These factors, among others, influence the ability of farmers to invest in soil-fertility on their farms.

But such household-level survey data only give a partial picture of how soil management fits into broader livelihood patterns. Access to assets interact in different ways, and are conditioned by a broad range of social, economic and demographic processes. In order to get a more complete picture of such

Table 2.5 *Socio-economic characteristics of different groups and case study farms*

	Highland		*Lowland*	
	Rich (N=13)	*Poor (N=37)*	*Rich (N=19)*	*Poor (N=31)*
Family size (N)	12.7	7.1	11.2	7.4
Land holding (ha)	1.8	0.6	5.2	3.5
Cattle owned (N)	9.1	1.0	6.6	1.0
Smallstock owned (N)	1.4	0.2	0.8	0.3
Off-farm cash income (Birr/year)	731	61	385	61

Source: Eyasu, 1997

dynamics, the following section offers a series of 6 case studies which look at how farmers of different levels of wealth and from different agro-ecological zones have managed their soils. Through the case studies, the broader contexts for soil-nutrient balances (discussed below) are provided.

Highland cases, case 1: Richer farmer

TC is a relatively rich, 36-year-old farmer from the highland part of Gununo area. He owns 3.5ha of land, 0.2ha of which he inherited from his father; the remainder was purchased. When a neighbour fell on hard times he purchased his highly fertile enset garden and *darkoa* garden. He now has accumulated a range of dispersed patches of high quality flat areas for farming. He owns 11 cattle, including 4 oxen. He keeps only 5 cattle at home, as the remainder are out at poorer farmers' houses on loan arrangements (*hara*), particularly with farmers in the lowlands. These cattle are fattened up for sale at local markets. This cash income combines with trade in consumer goods to allow him to purchase 150kg of DAP fertilizer each year. Due to his relatively young age he has only 6 young children, so he employs labour at key periods. Access to cash is therefore critical to the success of his farming enterprise. Through the cash he raised from trading, he bought more land and cattle. These renewed the existing land (particularly through manuring and the building of soil bunds) and he was able to buy more quality land in different locations, allowing crop diversification and the expansion of intensive gardening in the enriched garden areas. In terms of nutrient budgets, his garden areas (enset, *darkoa* maize and taro) are all either positive or approximately in balance for N and P. However, his more distant *shoka* field shows a significant negative balance for N, due to erosion and leaching losses of the fertilizer applied.

Case 2: Mid-wealth group farmer

BC is 55 years old and farms a plot of just over 1ha. He owns 7 cattle which are all resident on the farm. He supplements his income through a sharecropping arrangement on a small portion of his outfield. In addition, the family gains income from his wife's trading activities. With the assistance of a number of sons and daughters, BC has been able to invest in the improvement of the soils in his farm. In addition, he has capitalised on past investment in soils by tenant farmers on his land. His major strategy is to increase the area of garden by pushing the frontier of fertile soil out from the house and into the infertile, *shoka* field. Following a period of two or three years of intensive management and high levels of organic-matter incorporation, as well as investment in soil conservation measures, the plot becomes part of the garden area (*darkoa*). Over the last decade, BC has expanded the area of fertile garden significantly, leaving less than half his land as *shoka* outfield. The aim, he says, is to eliminate this completely and transform the whole area into a garden. However, manure supplies are currently insufficient and he must purchase fertilizer for the maize planted in the *shoka* area which is applied in small amounts as a top dressing. In terms of nutrient budgets, the garden areas are all positive, while the *shoka* area shows a negative balance for nitrogen.

Case 3: Poorer farmer

HS is 60 years old and farms a small plot of 0.25ha in the highlands. His father had a large area of land, which was split among 6 brothers on his death. In the past he owned a number of cattle, but due to the death of family members cattle were sold for funeral costs, and during periods of extreme food shortage. These days he owns no cattle and so must rely on cattle loaned to him for fattening. In order to attract loan arrangements, he maintains a large grass area at the lowest point of the farm which he grows for stall animal feeding. Through the intensive feeding and manure-management system he has developed, he manages to produce a large amount of manure from each animal held. This is a critical input into his soils. Due to a lack of other sources of income he purchases only about 2kg of fertilizer in a season which he adds very selectively. Otherwise he must rely on organic sources, including bulked-up manure and leaf litter. All labour on the small farm is family labour. Despite his family's best attempts, the enset area has been contracting due to premature harvesting because of food shortage. He has recently realized he cannot sustain the required level of manure input for the perennial and root crops and is now increasingly releasing land for annual crops. These are produced with an intensive, organic-matter-based gardening system, with high per-unit-area productivity. In terms of nutrient budgets, the enset garden and *shoka* field area show a negative balance for nitrogen, while the *taro* and *darkoa* garden area, where much of the management effort is invested, is approximately in balance.

Lowland cases, case 4: Richer farmer

BA and his family established their farm in the Bele settlement area in 1975. At that time, the newly-cleared land was poor. They did not have any gardens as they had done in the highlands from where they came. However, WADU was there and they provided fertilizer, so when the garden area was small, crops could be reaped. As a farmer leader under the WADU system, he gained easy access to fertilizer. Over time, he has increased the area of garden bit by bit. Around four buckets of manure and household refuse are produced each week and from October this is added to the garden regularly. Ploughing during

October helps incorporation and prevents the termites from removing all the valuable organic matter. Fortunately he was not moved as part of villagization, and the gardens have grown. When the fertilizer was stopped when WADU left, they had to rely on their gardens for maize production. The outfields are now for maize, sorghum, sweet potato, haricot beans and some teff. Only a small portion receives about 100kg of fertilizer, which is purchased at the local market. Since he has around 10 cattle, sufficient manure is produced. In the past, before trypanosomiasis really struck hard in the area, he had over 50 cattle. Nevertheless, in 1995 around 6 tonnes of manure was transferred to the *darkoa* field and around 0.2ha of new *darkoa* land was created through intensive manuring and hoeing. Overall, N is more or less in balance in the *darkoa* areas, while P balances are positive. On the *shoka* field, by contrast, large negative N balances are observed, although P balances remain positive.

Case 5: Medium-wealth group farmer

AA is 56 years old and farms a 5ha area, of which the fertile *darkoa* maize garden and taro root field account for 0.5ha and 0.25ha respectively. The family has 5 cattle, plus 2 goats, all of which are owned by the family. Due to a lack of 'seed money' no-one in the family is involved in off-farm trading or other activities, making the farm the key source of livelihood. Since settling here as a young man, AA has improved the land quality of the farm, particularly the garden areas which now are quite fertile. In 1995 AA purchased 75kg of DAP fertilizer which was applied to the maize in the *shoka* field. In addition around 4 tonnes of manure was transported to the *darkoa* and *taro* plots. In addition, leaf litter and household waste was applied to the garden area. In terms of nutrient balances, the garden plots show positive N and P balances, while the *shoka* field has a strongly negative N balance.

Case 6: Poorer farmer

MA has a small plot of land of around 5ha. Five years ago he sold around 2.5ha to another farmer, as he had to purchase food for the family. During the villagization period, they were forced to move. However, he continued to farm his old plot and managed to avoid the soil deteriorating. When the new government came, he returned to his old site, and reestablished his home, this time in the area of the poorest soil. Today, his main field is where his old home was and it is therefore quite fertile. The new garden around the home is being enriched through intensive hoeing and the addition of grass, leaf litter, household waste and manure. The manure is derived from his own small flock of goats and the manure from cattle he holds for fattening. The intensive feeding regime means that relatively large quantities of manure are produced. No inorganic fertilizer is used at all because of the cost, although some was applied recently to a portion of his outfield that he let out under a sharecropping arrangement to the church. Home production combined with sharecropping and rearing are the basis of the household economy. Due to the recurrent sickness of his wife, she is unable to trade. As a result, crop residues and grasses are returned to the field, rather than sold as many others do. In terms of nutrient balances, all fields show negative N balances, although P levels are more or less in equilibrium.

Case studies: understanding livelihood dynamics

These cases illustrate the different ways in which access to land, labour, livestock and capital is gained, and the way that this access is mediated by a range of institutions. For example, in the highlands, access to land for the richer farmer has been reliant on access to cash income from trading. This has, in turn, meant he has been able to enter the emergent land market and purchase highly fertile land parcels, previously enriched by others. Others must be reliant on direct relationships with tenant farmers to work their land, as in the case of the medium-wealth group farmer in the highlands, or sharecropping systems to get land at all, as is the case with many poorer households. Labour access is similarly varied, ranging from reliance solely on household labour, to fosterage relationships, to direct hiring of labourers for particular tasks.

Labour for soil-fertility management is also mediated by social relations within the household. Thus gender as well as age relations affect whether men or women carry out particular tasks. In the Wolayta case, most field-based soil-fertility management tasks are carried out by men, particularly younger men if they are around, while women are more involved in activities more associated with the home, including manure management, composting and transport of wastes to the nearby garden area.

The cases also show how livestock access is mediated through a variety of institutions, including a variety of loaning arrangements, particularly for cattle. These are, it appears, particularly important for poorer households who, without such a loan, would have no access to cattle manure at all. Finally, access to cash for the purchase of inputs is also highly differentiated, with off-farm income sources, derived from a major trading business in the case of the highland rich male farmer, or through petty trading by women in local markets in the case of the medium wealth household from the highlands.

Impacts on farm- and plot-level nutrient balances

How do these various soil-fertility management practices and socio-economic contexts translate into physical changes in different field plots? This question was investigated through a study of the N and P budgets in four different farm components – the enset garden, the maize *darkoa* garden, the *taro* field and the maize *shoka* outfield – for 8 farmers in the highlands and the lowlands, including those case studies highlighted above.

For each case study farm, a simple nutrient-budget model was compiled for the whole year, across both seasons. This involved measuring or estimating all the inputs and outputs for each farm sub-component. The balance

(NUTBAL) for both nitrogen (N) and phosphorous (P) is estimated according the equation:

$$NUTBAL = IN1 + IN2a,b + IN3 + IN4 - OUT1 - OUT2 - OUT3 - OUT4$$

where *IN1* = mineral fertilizer; *IN2a* = manure/household wastes; *IN2b* = leaf litter; *IN3* = atmospheric deposition; *IN4* = biological nitrogen fixation; *OUT1* = crop yield; *OUT2* = crop residue removal; *OUT3* = leaching and denitrification; *OUT4* = erosion.

With the exception of atmospheric deposition, fixation, leaching, denitrification and erosion which were estimated, all of the other parameters were directly measured on a regular basis throughout the year (Eyasu, 1997; Eyasu et al, 1998). While there are many limitations of such an exercise (see Chapter 1), and the final figures should not be taken too seriously, the broad patterns are of interest, especially when the results are disaggregated by the type of farmer and the agro-ecological. Table 2.6 provides a summary of the results of the nutrient-balance analysis for N and P by farm sub-component.

In terms of nitrogen balances, the study found that in the highlands the garden areas (enset, *darkoa* and taro) were overall approximately in balance, with poorer farmers showing relatively small negative balances in the enset and *darkoa* areas due to less manure availability. By contrast, the *shoka* outfields showed consistently large negative N balances due to the outputs of harvested crops not being compensated for by fertility inputs. This broad pattern is repeated in the lowland areas, although negative N balances are shown across all farm sub-components, again with poorer farmers having larger deficits per unit area than the richer farmers due to lack of resource access. P balances, by contrast, show a different pattern with positive balances across all farm sub-components and all socio-economic groups, except for small negative balances in the taro gardens and *shoka* fields of the poorer lowland farmers.

In contrast to the generalized data presented in much of the policy literature,[8] a highly differentiated pattern of nutrient depletion and accumulation is seen, which varies by both site and the socio-economic characteristics of the farmer. Not surprisingly, these patterns reflect differences in soil-fertility

Table 2.6 *Nutrient balances (N/kg/ha and P/kg/ha) by farm sub-component, disaggregated by socio-economic group and agro-ecological zone*

	Highland				*Lowland*			
	Rich		*Poor*		*Rich*		*Poor*	
	N	*P*	*N*	*P*	*N*	*P*	*N*	*P*
Enset garden	11.5	11.0	–12.0	5.5	–	–	–	–
Darkoa maize garden	–3.0	8.0	–4.5	4.0	–4.0	10.5	–24.0	3.0
Taro garden	4.0	10.5	–0.5	0.5	–7.0	8.5	–9.0	–2.6
Shoka maize outfield	–95.0	6.5	–54.0	3.0	–20.0	6.5	–40.5	–1.0

Table 2.7 *Soil amendment practices (percentage of households employing a certain practice)*

	Highland		Lowland	
	Rich (N=13)	Poor (N=37)	Rich (N=19)	Poor (N=31)
Fertilizer	100	62	100	74
Manure	100	86	100	74
Leaf litter	0	38	37	45
Compost	8	8	21	10
Termitaria	0	11	0	19

Source: Eyasu, 1997

management practices employed by different groups across the two zones. Table 2.7 reports the results of a sample survey of 100 households, which included the case study farms discussed above, on soil amendment practices.

Some broad patterns are evident. The richer farmers are basically reliant on a two-prong strategy of using purchased inorganic fertilizer and manure from livestock on-farm. As the case studies and the nutrient budget analyses show, these are applied to different areas of the farm, with manure being focused on the garden areas (enset, maize *darkoa*, and taro field) and inorganic fertilizer being used in the *shoka* field. By contrast, poorer farmers make use of a much more diverse range of fertility inputs, including a range of organic sources like leaf litter, compost and termitaria, to complement their limited access to inorganic fertilizer and manure. The total amount of these materials is limited, and therefore their application is focused on the garden areas, with few inputs at all reaching the *shoka* outfields. In terms of inorganic DAP fertilizer, only 34kg and 33kg of DAP is applied to poorer farmers' fields on average, compared to 151kg and 85kg in the richer farmers' fields in the highlands and lowlands respectively.

What are the implications of these results? Does it matter that outputs exceed inputs in many of the farm components? What are the longer-term implications for the agricultural system? These questions are, of course, not easy to answer. However, a rough indication can be given if we look at the nitrogen and available phosphorous in the soil. Soil samples taken from the different farm components show that there are high levels (0.20–0.32 per cent) of available N in all the highland soils, but relatively low levels (0.10–0.13 per cent) in the lowland soils, suggesting that N removal in the lowlands, particularly in the *shoka* areas, is potentially a serious problem. Available P is also low in the lowlands (3.75–7.46ppm [parts per million]) in both the garden and *shoka* areas, but is higher in the highland gardens, especially in the enset gardens (over 40ppm) (Eyasu, 1997). Within the highlands, the *shoka* area is the only place where response to addition of P is probable, while in the lowlands response to both N and P is possible, given the soil test results (Young, 1980, p291).

With the type of nutrient removals or additions suggested by the monitoring results (Table 2.7), the continued loss of N from the *shoka* area in the highlands across both socio-economic groups is undoubtedly a cause for concern, as is the loss of N in all lowland sites. In these situations real yield losses can be expected now or in the near future unless compensated for. However, the P situation looks much more promising, with positive balances in almost all areas. With such continued inputs the P levels of the soil will be built up and the current low levels being experienced in the lowlands and the *shoka* fields in the highlands may be offset, assuming immobilization is prevented.

However, even if trends are not inexorably negative, nor likely to result in severe constraints to production levels in the near future in most sites, this does not imply that additions of nutrients are not required to boost crop production. For example, fertilizer experiments in outfields in the area have long shown that additions of P result in yield improvements. Indeed, as we have seen, the major component of WADU's agricultural improvement programme was based on the promotion of DAP fertilizer. Crop monitoring results show the results, with significant boosts in yields of maize and teff in *shoka* outfields experienced during this period as a result of increased levels of inputs (Figure 2.4). Since the withdrawal of the subsidized programme in 1981, yield levels have returned to the previous level of around 1000–1500kg/ha in the highlands and around 750–1000kg/ha in the lowlands.

Since the end of the WADU era, therefore, farmers have had to rely much more on their own on-farm resources, supplemented by inorganic fertilizers if cash or credit is available. Even for the richer farmers in the highland area, application levels are very low (see above), with the result that present yield

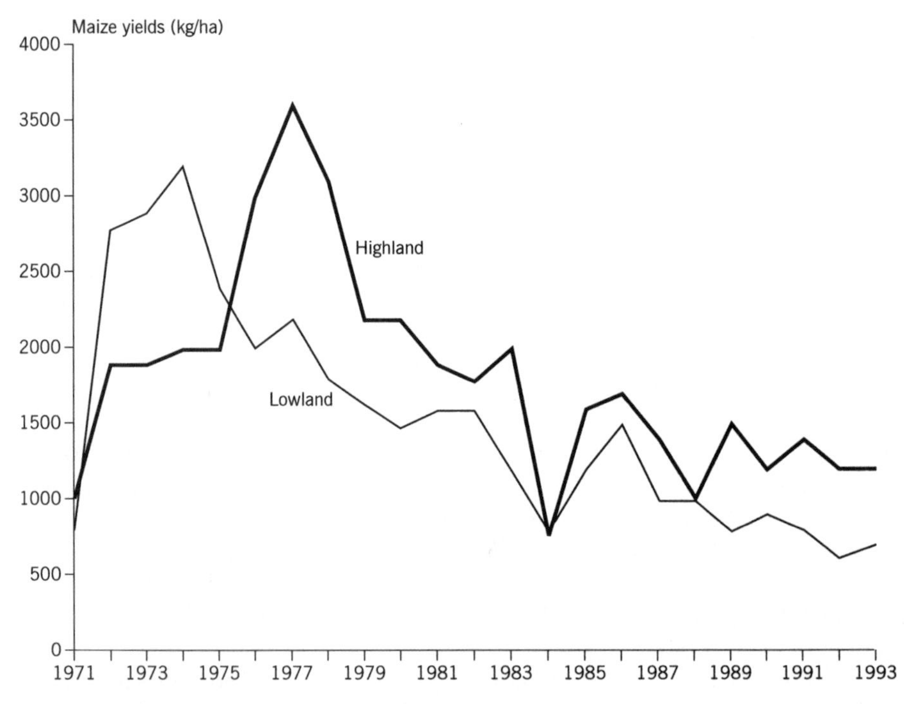

Figure 2.4 *Maize yields on farmers' outfields in Kindo Koisha, 1971–1993, highlands and lowlands*

levels for maize and other crops are low and appear to be stagnant. Figure 2.4 shows, for instance, that, despite increases in rainfall occurring in recent times, yield levels have not responded, suggesting the possibility of fertility limitations. The pattern observed in Wolayta appears to be reflected more broadly in Ethiopia, with no significant yield improvements being shown across a range of crops since 1980. Fertilizer use rates are on average 35kg/ha per year (95kg/ha per year for fertilizer users). While this is a considerable increase over 15–20 years ago, this remains low given the challenges of food production in Ethiopia (Mulat et al, 1997).

CONCLUSION

So what are the implications of these findings for food security and livelihoods, and the broader policy of agricultural-led growth in Ethiopia? Declining land areas, increasing populations and reductions in input use suggest rather a dismal prognosis for the future. A number of national surveys indicate an increasing level of poverty in a number of areas, including the south (Dercon and Krishnan, 1996a, b; Bevan and Bereket, 1996; Bevan, 1997). Some of this pessimism is shared by local commentators. Ato Altaye Anaro from Fachana PA observes:

> *In my father's time there were few people and most of the land was for grazing. Now there are many people with small fields, all cultivated. With more homes, the* darkoa *areas are expanding and the fertility of the soils is increasing. However, the* shoka *land is suffering from lack of fertility. Now people must crop every space and plant in every season. The future of farming in this area is not good. There are too many people. They will have to go elsewhere to large farms and work. If they have to live here, in 10 years time, people will have to build houses over houses, like we heard of the houses in big towns.* (Interview, 1996)

However, in contrast to the generalized statements about poverty and land degradation which dominate the policy debate and research literature, Ato Altaye points to a number of key nuances, highlighting a contrasting set of apparently contradictory dynamics of change. Population increase and contracting land areas, it seems, have prompted increased investment in garden areas (*darkoa*), through intensive manuring, hoe cultivation and careful management. This process of intensification is associated with significant increases in soil-fertility (as reflected in the nutrient balance studies reported above) in certain patches. With more homes and more people, there are more gardens and more investment in soil-fertility. This is not the pessimistic 'downward spiral' of environmental decline and associated poverty which so many commentators diagnose.

The dilemma remains that, despite such investments in soil improvements, this is insufficient to provide enough food for many in a considerable proportion of years. It is clear that a parallel set of investments are required in the *shoka* fields where, with declining fertilizer inputs, yields are stagnant and remain low (Figure 2.4). Increasing fertilizer inputs to crop production in the *shoka* fields remains a major practical and policy challenge, requiring attention to credit institutions, infrastructure development and input supply market reform. The current system results in significant costs to fertilizer users, resulting in declining profitability of fertilizer use. If such trends continue, the prospects for boosting production on the *shoka* fields remains remote, as the manuring strategy of the *darkoa* gardens is reaching its limits due to the limited availability of manure, particularly for the poor.

While much of the policy focus is appropriately on increasing supply of fertility inputs, the challenges of increasing the efficiency of use of available materials should not be overlooked. Many have to make do with very limited quantities of manure – perhaps from a single animal – and, due to cash limitations and the limits of credit support, purchase of fertilizer is limited to a few cupfulls. Increasing the efficiency of use of such valuable nutrient resources is therefore a major challenge. On-farm trial work with farmers has begun to look at such issues. Participatory experimentation has begun to investigate ways of increasing the nutrient content of manures and other biomass resources through composting; approaches to reducing nutrient losses through volatilization and leaching via more effective manure and urine management techniques; strategies of application which allow a more effective distribution of nutrient resources across a farm; and fertilizer placement and timing techniques which maximize plant uptake (see Chapter 5).

Thus the challenge for Wolayta agriculture and soil-fertility management is to envisage a differentiated approach to tackling the issue. A simple 'technical fix' – whether in the form of a fertilizer-based package programme or a more complex mixed strategy based on organic and inorganic inputs – is clearly insufficient. Different pathways of environmental and livelihood change are evident in different agro-ecological zones, in different parts of the farm landscape, and across different farm families with different resource endowments.

At the farm level, there is a clear distinction between processes of intensification in the garden (*darkoa*) and outfield (*shoka*) areas, where different technical and managerial solutions are required. Differentiation at this level is important because, as the WADU experience and more recently the SG–2000 package approach have shown, a concentration on a capital-intensive fertilizer-based approach is only appropriate in the long term for a relatively small sector of the population. And, even for those who can afford such a strategy, the danger of undermining the root crop and enset part of the system is apparent if the continued, labour-intensive investments are withdrawn.

As the case study research has shown, livestock are essential for this component of the Wolayta farming system, and the linkages between garden-based crop production and livestock management is vital for productivity and sustainability. While livestock populations have increased overall holdings have declined, making manure a highly valued resource. The former strategy of harvesting nutrients from lowland grazing areas through the movement of animals between the highlands and lowlands is increasingly constrained, especially because of increasing tsetse challenge and trypanosomiasis risk in the lowlands. This makes the challenges of efficient grazing and fodder management in the highlands and livestock disease control in the lowlands especially important.

But, as previous sections of this chapter have shown, agriculture and livestock production is only one component of the overall livelihood system. With decreasing land and livestock holdings, the option of relying solely on agriculture is out of the question for most. Off-farm activities and migration have long been part of people's livelihoods in Wolayta (Carswell et al, 2000), but these strategies will become increasingly significant in the future, both as separate sources of income, but also as means of securing investments in the aricultural system. Thus soil-fertility management must be looked at in this broader context, as one element of a wider livelihood system. Generalized statements and broad prognoses about environmental decline, accompanied by a technically-focused set of solutions, miss key issues of diversity and dynamics which must be central to any future strategy for development.

Chapter 3

Seizing New Opportunities: Soil-fertility Management and Diverse Livelihoods in Mali

Ibrahim Dembélé, Loes Kater, Daouda Koné, Yenizie Koné, Boutout Ly and Allaye Macinanke[1]

Introduction

Soil-fertility is critical to an agricultural economy such as Mali. Whether in the dryland farming areas in the centre of the country, the cotton zone in the south, or the irrigated rice areas of the Office du Niger, much concern is expressed about the potential for land degradation as a result of soil-fertility decline. Thus, for example, the recently prepared national action programme to combat desertification notes the:

> *accelerating rate of soil erosion… and alarming desertification encroachment … [and the need] to impel a change of mentality and behaviour and ensure an efficient participation of the population and the various actors involved in the elaboration and implementation of programs on environmental protection/management.* (Government of Mali, 1998, piii)

Studies carried out in association with the Soil-fertility Initiative speak of the:

> *worrying growth in soil degradation… the over-use of resources… stagnant or falling yields … growing food insecurity…. demographic growth and the irrational use of natural resources.* (Gakou et al, 1996, p2)

But what do these generalized proclamations look like on the ground? How significant are they in different regions? What implications do they have for farmers?

This chapter is concerned with looking at what happens in farmers' fields across a diverse range of Malian farming systems. Case studies from rice, cotton and rainfed cereal cropping zones are used to explore how different groups of farmers manage their soils and address land degradation and soil-fertility decline. The chapter starts with an overview of the Mali setting, introducing policy issues, institutional structures and an outline of the technical debate about agriculture and environment. It goes on to describe the main farming systems and the economic and institutional changes which have affected them. This is followed by an introduction to the research sites and their broad characteristics. We then turn to a discussion of how farmers perceive soils and try to supplement them with various fertilizing materials, before examining the evidence from nutrient balances regarding the effectiveness of such strategies for different farmers and settings. It concludes by identifying implications of these findings for policy in terms of research and extension approaches, opportunities for broader livelihood diversification, and changes to macro-level institutional and economic conditions.

THE MALIAN CONTEXT

Mali is a large landlocked country in the heart of West Africa, covering 1.27 million km^2, half of which receives less than 200mm of rainfall a year (see Figure 3.1). The single rainy season is followed by 7 or 8 months without rain, when the Harmattan winds blow. Rainfall is highly variable from year to year and from place to place, with variability increasing towards the north. The country is transected by two major river basins, the Niger and Senegal, and their various tributaries. Patterns of population density reflect rainfall to a large extent, with levels as low as 1–2 per km^2 in the northern regions, rising to 50 per km^2 in the south.

Formerly part of French West Africa, Mali became independent in 1960. A coup d'état in 1968 brought a military government to power which was finally toppled in 1991. A democratic multi-party system was established in 1992 under which two general elections have been held. Per capita income in Mali is estimated at US$240, which, in combination with poor levels of social provision, gives Mali a ranking of 171 out of 175 countries in the world according to the UNDP's human development index (UNDP, 1999). Mali is also considered a highly-indebted poor country which means access to debt relief measures in return for continued administrative and fiscal reform.

Questions relating to rainfall and risks of drought form an important background to agricultural and environmental issues in Mali, as elsewhere in the Sahel. The severe droughts of 1973–1974 and 1984 had major impacts not only on human suffering, the national economy and impoverishment, but also at a policy level. For example, the great Sahelian drought of 1973–1974 gave added emphasis to the need to develop and extend the irrigated area. It also led to added emphasis on soil and water conservation activities, natural resource management and *gestion de terroir*[2] approaches. The drought years also brought a great increase in donor funding to Mali. Over the period 1985–1995, one-third of GDP was contributed by aid, though this share is now declining. Sales of cotton are of great importance, with Mali currently the second largest cotton producer in Africa. The future of the cotton zone is therefore critical to the Malian economy, making the CMDT (Compagnie

Malienne pour le Développement des Fibres Textiles, the company which manages the production and marketing of cotton) a key economic and political player in the country.

In addition, the droughts accelerated the reform of state structures and ambitions, moving from the socialist model pursued in the years following independence towards the pro-market model being followed by the current government. Mali was one of the first African countries to engage in a programme of structural adjustment measures following a World Bank mission in 1982. Structural adjustment in Mali has involved a broad range of measures, which include cutbacks in government personnel and extension officers, sale of state enterprises, liberalization of cereal markets, abolition of fertilizer subsidies, imposition of charges for services such as veterinary care, and fiscal reform. Together with the other members of the Union Economique et Monétaire de l'Afrique de l'Ouest (UEMOA), Mali devalued its currency, the CFA franc, by 50 per cent in January 1994, a measure strongly pushed by the IMF and World Bank, with significant impacts on the farm sector. Cereals now face better markets in urban areas since food imports have become more expensive. The impact on cotton has also been broadly positive for most farmers, although its cultivation is dependent on external inputs which have become more costly. Equally, livestock producers have been able to take advantage of improved market conditions in the richer coastal states, which formerly relied on much imported frozen meat (Kébé et al, 1997).

The period following the droughts also witnessed a mushrooming of NGOs, initially international organizations bringing short-term relief, which then moved towards longer-term development activities. Over time, a parallel grouping of Malian NGOs has also grown up, many as partners of international NGOs, a few as more opportunistic attempts to take advantage of the vogue amongst donors to work with and fund 'civil society' organizations.

While Mali may lack a dense web of formal organizations beyond the NGO sector, Malian society is rich in institutions and informal structures within which people live their lives. Foremost amongst these is the household which provides a very important economic and social framework which plans and manages resources, not only in the farming sector but also in urban areas and trading networks. Village-level structures continue to carry out certain functions, such as regulating access to land and water, and mediating in cases of conflict. In addition, inter-household linkages and those between people in the same age sets provide channels of mutual help and support in times of need (Pollet and Winter, 1971; Meillassoux 1975; Brock and Coulibaly, 1999).

Building on such traditional institutions during the early years of independence, the state tried to establish a system of co-operatives throughout the country. They did not function well, and have largely disappeared, with *Associations Villageoises* taking their place in cash-crop areas (Beaudoux and Nieuwkerk, 1985; Degnbol, 1999). Their purpose has been to help market harvests, provide inputs, and manage credit. In the cotton areas, dissatisfaction with the operations of the cotton marketing company led in 1992 to farmers creating a union of cotton producers, known as the Syndicat des Producteurs de Coton et de Vivriers (SYCOV). This union has demonstrated its capacity to mobilize its membership and negotiate better terms regarding prices with the cotton company (Bingen, 1997; Koné, 1997; Degnbol, 1999). Elsewhere, farmer organizations are much less common. Equally, herders' associations are few and far between and very often a construct of a donor project, rather than constituting a body with real legitimacy and representation of local people.

In the last few years, Mali has been preparing for the establishment of a new decentralised system of administration which, if effective, will fundamentally transform the political, administrative and organizational landscape of rural areas (Sy, 1998). Following elections held in mid-1999, new local authorities, the rural communes, were established. These rural communes are meant to have considerable powers to mobilize resources both within their area and from external sources for various development activities. They are also responsible for managing and allocating rights over natural resources within the commune area, in consultation with village communities in accordance with the new *Code domaniale et foncier* which aims to clarify the relationship between statutory law and customary practice. However, some concerns have been raised regarding the likely areas of conflict and tension between customary structures at village level and the new rural communes, particularly over management of common property resources (Dème, 1998; Hilhorst and Coulibaly, 1999).

AGRICULTURAL AND ENVIRONMENTAL ISSUES IN MALI

It is against this general background that more technical agricultural and environmental issues in Mali are being discussed within the context of government priorities and donor concerns. In terms of agricultural production, the availability of both soil moisture and nutrients is critically important (see Chapter 1). Water limitation is key, especially in drier parts of the country, and has been becoming more acute with declining rainfall levels. A comparison of rainfall data from 1931 to 1960 with the subsequent thirty year period from 1961 to 1990 shows the latter period received 20–25 per cent less rainfall (Hulme, 1992). Expressed in another way, rainfall bands have in effect shifted southwards by 100–250km since 1960. This decline in rainfall is perceived by farmers as being responsible for many of the difficulties they face. However, since 1984, there has been no worsening of rainfall levels, and certain years, such as 1988 and 1997, have been comparatively wet. Nevertheless, there is considerable uncertainty about future rainfall trends, impacts of global warming and the influence of other climatic events such as El Niño.[3]

Rainfall levels interact with soil-fertility to affect the potential productivity of land in important ways, especially given the poor quality of the heavily weathered and ancient soils found in much of Mali. A major ecological study – the *Production Primaire au Sahel* (PPS) programme – was carried out in the 1970s, in different sites in Mali. This identified soil nutrients as being the key limiting factor in large parts of the country, even in areas with relatively low rainfall (Penning de Vries and Djiteye, 1982; Breman and Traoré, 1987). Such a focus on soil nutrients was further developed by the *Production Soudano-Sahelienne* (PSS) programme which concluded by emphasizing the critical need for increased use of external inputs, including inorganic fertilizer and animal feed, without which it would be impossible to meet the food requirements of a growing population (van Keulen and Breman 1990; Breman and Sissoko, 1998). These findings echo work from across West Africa, which highlights soil-fertility as a key issue (Bationo et al, 1998; Pichot et al, 1981; Pieri, 1989; Floret and Serpantié, 1993; Sebillotte, 1989; de Ridder and van Keulen, 1990). Alongside the production ecology modelling approach of the PPS study, long-term experiments and field trials of various sorts have been carried out in a range of sites. These have shown how important soil-fertility management is to the sustainability of production systems (Bosma et al, 1993; Camara, 1995; de Vries and Prost, 1994; Doumbia, 1998; Dureau et al, 1994).

A common set of findings has resulted from these studies which is repeated in the policy statements cited earlier, as well as other strategy documents (eg Gakou et al, 1996; Government of Mali, 1998). The main lines of the argument are that significant applications of inorganic fertilizer will be required to improve agricultural production in Mali, and the Sahel more broadly. This is due to inappropriate cultivation methods, the reduction of fallowing and the lack of manure and organic biomass to make good the deficits caused by 'nutrient mining'. Significant efforts must also be invested in reducing levels of erosion and stemming broader processes of desertification and land degradation.[4] In the Malian context, particular concern is focused on the cash crop zones, with lower rainfall and soil-fertility decline being seen as major factors accounting for the decline in cotton yields in the older cotton areas in Mali-Sud, although broader problems of land degradation are assumed to cover all zones.

A number of studies have attempted to estimate the economic costs of current farming practices. Drawing on aggregate nutrient budget estimates for the cotton zone, van der Pol (1992), for example, argues that soil nutrient depletion constitutes around 40 per cent of income gained by farmers in southern Mali, yet is being ignored. In a similar vein, Bishop and Allen (1989) calculate that soil erosion is causing losses equivalent to 12–17 per cent of Mali's agricultural GDP each year. Despite the fragility of their assumptions and the questionable basis of some of the underlying data,[5] such statistics have captured the attention of policy-makers and donors and are subsequently referred to repeatedly (see Chapter 1).

Agricultural development strategies have pursued two major lines of activity. The first centres on cash crop areas, particularly the cotton zone in southern Mali, where a combination of extension support, including a major campaign to encourage soil conservation,[6] and the encouragement of increased fertilizer use are central to government policy to ensure continued cotton production. Considerable government and donor investment has been made over the last few years through support to the CMDT for agricultural research, extension, input supply, credit, and marketing issues. This continues to remain a high priority, with soil-fertility issues being seen as intimately linked to the viability of the cotton economy.

A second focus has followed a more locally-focused approach, centred often on the *gestion de terroir* approach. Such approaches cover both cash crop and dryland cropping zones. Thus, for example, a large number of NGOs are involved in support to small-scale soil and water conservation projects, aimed at adapting and strengthening traditional measures to tackle erosion (Reij et al, 1996). Such projects also address broader environmental issues, such as management of the village's lands, tree planting, the introduction of improved wood stoves and access to credit. Increasingly, it is recognized that the conservation of soil and harvesting of rainfall run-off need to be combined with soil-fertility measures. Many NGO projects tend to maintain very limited contact with the Malian agricultural research sector, and rely on a more 'hands-on' approach, using a range of international networks and contacts through which to gain access to information and new ideas.

Such intervention oriented approaches are linked, in various ways, to a number of overarching policy initiatives emerging from the international environmental arena. For example, the Malian government recently held a roundtable consultation with donors on the newly formulated National Action Programme to Combat Desertification (Government of Mali, 1998). Attempts have been made to link this process to Mali's parallel commitment to the World Bank, to design and implement a National Environmental Action Plan (NEAP). At the same time, steps have been taken to get Malian engagement with the Soil-fertility Initiative (SFI), an international venture being led by the World Bank and FAO. A Soil-fertility Action Plan has been proposed which would be developed through a series of participatory workshops at regional levels, to examine ways of improving the availability of farm inputs – especially mineral fertilizer – and to establish a fertilizer factory to make use of local rock phosphate supplies (FAO, 1998; Truong, 1998).

The rest of this chapter is concerned with exploring the diversity and dynamics of soil-fertility management at a local level through four detailed village-based case studies in contrasting agro-ecological zones. Through a detailed examination of social, economic and biophysical processes, combined with insights from farmers' own commentaries on the issues they face, the following sections will explore whether the broader prognoses discussed above are appropriate, or whether alternative options for farm- and policy-level intervention are possible.

FARMING SYSTEMS IN MALI

Agriculture in Mali can be sub-divided into two major zones. On the one hand are those areas characterized by a dominant cash crop, which include the Office du Niger irrigated rice scheme, and the cotton zone in southern Mali. Farmers here invest considerable amounts in the purchase of mineral fertilizer and other agricultural inputs. However, even in the cotton zone, cereals often take up more than half the cultivated area, being grown in rotation with cotton. On the other hand are the zones where no major cash crop is grown, though grain sales may be an important source of income in years of good

Figure 3.1 *Map of Mali showing the case study sites*

rainfall. These zones, in the centre and more northerly part of the country, are characterized by extensive rainfed agro-pastoral systems producing livestock, millet, sorghum and cowpea. Here, markets are often distant and access to inputs very limited, with farmers still relying to a large extent on animal dung and fallowing to restore soil-fertility.

This distinction between cash crop and dryland cereals is evident in the priority given to each area, in terms of access to resources, research and extension activity and donor programmes. Such differences in priority are, in turn, justified on the basis of their respective contributions to the national economy and prospects for future growth. Of the four study villages, Tissana and M'Péresso are in the irrigated rice and cotton cash crop systems respectively, while Dilaba and Siguiné represent rainfed cereal cropping areas. The four villages are situated on a north–south gradient, as seen in Figure 3.1.

The Office du Niger irrigation scheme (research site: Tissana)

The Office du Niger irrigation scheme was started in the 1930s by the French colonial administration to produce cotton and currently covers an area of 50,000ha. Initially, few farmers from neighbouring areas were interested in moving to the scheme and the French had to bring settlers from as far afield as Burkina Faso's densely-populated Mossi plateau (Schreyger, 1984; Magasa, 1978). Recently, however, more and more people have moved to the Office as the benefits of irrigated production have become more evident; as a result the Office now covers 150 villages, with a total population of around 150,000 people (Jamin et al, 1995).

Irrigated cotton cultivation was never very successful and was gradually replaced with rice farming from the 1960s onwards, which also suited the government's strong interest in providing sufficient food grains for growing urban centres. Until the 1980s, the Office used to manage water, control access to plots and dictate the size of landholdings and production methods, constituting a 'state within a state' in those regions where it operated. The Office was also responsible for input supply, processing, marketing, extension and credit. Sales of rice were tightly controlled to ensure farmers only sold their rice to the Office, and all transport movements into and out of the rice growing area were monitored, to avoid illicit sales of rice. However, low rice prices and low levels of productivity led to rising indebtedness and default on loans so that, by the early 1980s, the Office was in serious financial difficulty.

A major rehabilitation of the scheme was carried out during the 1980s with support from several donors (levelling of land, digging out canals, installing more effective water control structures). This was followed by the introduction of new technologies such as improved rice varieties, production and planting of rice seedlings instead of broadcasting seed, recommendation of higher rates for use of mineral fertilizer, and the spread of double cropping. These have resulted in a three-fold increase in rice yields from an average of 1500kg/ha in the 1970s to around 5000kg/ha in the 1990s (Jamin, 1995; Maiga et al, 1998).

The operations of the Office had long been seen by many donors as in need of reform, and in 1991 important changes were made in the mandate and functions of the organization. Its responsibilities now focus on water management, maintenance of the scheme and land allotment, having withdrawn from other tasks such as extension and marketing. Credit provision had already been shifted to banks and farmer organizations (*Associations Villageoises*, AV) from the late 1980s. The rice market was liberalised in 1991 and controls over transport of rice lifted. The terms of trade further improved when the CFA franc was devalued in 1994,[7] with farmers gaining an estimated average increase in margin per hectare of 17–30 per cent (Maiga et al, 1998). However, the rather abrupt retreat from rice marketing by the Office has caused a number of problems. Farmers' lack of knowledge and inexperience with marketing caused them financial losses following liberalization, as they were often cheated when buying inputs or selling their produce. Credit systems also partly collapsed after the withdrawal of the Office, and village organizations have often not been strong enough to deal with their new responsibilities. In some villages all farmers lost access to formal credit, as a result of the AV falling into debt. In other villages while the AV may still function, those farmers who have accumulated debts cannot make use of its services. Recently new funding provision has been made available granting a reprieve to those AVs which had fallen badly into debt. But there are still a number of farmers who must rely instead on traders for access to inputs at planting time in exchange for payment in kind after harvest.

Access to land is still regulated by the Office. Farmers have a lease, but can be evicted if they do not comply with a series of regulations regarding how the plot is used and managed, or if they sub-let. Since there is considerable demand for irrigated land but little land available within the scheme, there are now plans to extend the area, and rehabilitate some of the less productive parts. There is also an informal market for both renting and selling leases despite this being technically illegal.

In the past, farmers in the Office relied very heavily on mineral fertilizer, but are now increasing the use of organic fertilizers. This is linked particularly to diversification by farmers into dry season vegetable production, which has expanded greatly since 1994. Farmers now grow a wide range of vegetable crops – onions, tomatoes, potatoes, lettuce and many others – both for sale in the market and under contract to commercial buyers.

Farmers usually have some fields outside the scheme where they cultivate rainfed sorghum and millet. They also have cattle and other livestock which in most cases are looked after by others, typically Fulani herders, and grazed outside the scheme for most of the year. The Office constitutes an important source of dry season grazing for local and visiting herds, which feed on the rice stubble during the dry season where land is not double cropped. However, this is causing increasing problems due to damage by herds to crops and canal banks, and increased competition over harvest residues, which can be used either as valuable dry season grazing for herds or for recycling on-farm.

The village of Tissana was chosen as a research site representative of the irrigated rice zone. Tissana is in the Office du Niger and mostly depends on the cultivation of rice. Fallow land in the area around Tissana is limited and is mainly confined to the *hors-casier* fields, which are situated within the boundaries of the Office but are not currently irrigated. Most of their cattle are put in the care of hired Fulani herders, who bring them to graze the rice stubble during the dry season. It was established in 1955 and lies close to the market town of Niono, which also houses much of the administrative and technical services running the Office du Niger.

Smallholder cotton farming in Mali-Sud (research site: M'Péresso)

Cotton production in southern Mali is carried out by smallholder farmers under rainfed conditions. While farmers traditionally grew a local cotton variety for domestic use, during the colonial period the French forced each village to produce a fixed quantity of cotton each year. This created much resentment and its memory led some villages in southern Mali to refuse to take up cotton production again until the 1980s.

Cotton was established in Mali as a cash crop in the 1920s and extension services created to help intensify production, through the introduction of animal traction, new agricultural practices and farmers' schools. The Compagnie Française pour le Développement des Textiles (CFDT), which was set up in 1947 to ensure a secure supply of cotton to France, exercised a monopoly over the export of cotton in a number of French-held African countries. CFDT activities in the Koutiala area of central Mali, where

M'Péresso is found, started in the 1950s. Cotton cultivation is now spreading into new areas in the sparsely settled lands of southwest Mali, and to former groundnut areas around Kayes and Kita in the far west of the country. The CFDT operations in Mali were nationalized in 1974, and replaced by the CMDT, a joint venture with the Malian government (which owns 60 per cent of the shares) and the CFDT which owns the remaining 40 per cent (Deveze, 1994; Fok, 1994).

Cotton production has been a profitable activity for all the parties involved, whether government, the CMDT or farmers. Mali has become the second-largest exporter in Africa with exports rising to a record of over 500,000 tonnes of seed cotton in the late 1990s. Cotton provides more than 40 per cent of the country's export earnings, while cotton oil and cottonseed cake are also important by-products, much of which is consumed locally. The CMDT has been involved in many aspects of research and extension, such as the training of village blacksmiths to produce and maintain plough equipment, the creation of AVs, soil conservation, and the management of village lands (Hijkoop et al, 1991).

Land is generally cultivated with cotton every two to three years in rotation with cereals. Cotton receives most of the mineral and organic fertilizer, while cereals benefit from their residual effects. Crops and livestock in southern Mali are closely integrated, with oxen used for land preparation, seeding and weeding, and organic manure is an important source of soil-fertility. Cotton stalks are used as bedding in the cattle *kraal*, and crop residues from cereals are grazed in the field and stored as fodder to be used at the end of the dry season. However, cereals are also of importance as are a range of other economic activities on and off the farm, such as fruit trees, trading and migration. A number of villagers in southern Mali have built up extensive holdings of land in neighbouring countries, such as coffee and cocoa plantations in Ivory Coast, and cotton farms in south-west Burkina Faso (Brock and Coulibaly, 1999).

A number of donors, led by the World Bank, have been pushing for privatization of the CMDT (Massou, 1998). The CMDT has already withdrawn rather abruptly from the provision of certain services, such as veterinary care, which it formerly assured. This has had damaging impacts on livestock health, with heavy losses to work oxen, which are of particular importance for preparation and weeding of cotton fields. The private sector in veterinary services is very small and weak, and has limited interest in meeting the needs of poorer farmers, especially in more distant areas. The CMDT also no longer provides credit for inputs and equipment but works through the banking system. However, it is still responsible for the purchase of inputs and the organization of the input delivery, transportation, processing and marketing of the cotton harvest. It also provides the main extension service in southern Mali in collaboration with the Regional Agricultural Research Centre in Sikasso. After considerable uncertainty about the future status of the CMDT, the World Bank announced in March 1999 that it would not seek its complete privatization, in return for greater transparency and efficiency in its mode of operation.

M'Péresso was taken as a case study site within the CMDT zone, and is representative of villages with a long-established history of cotton-growing, having taken it up in the 1950s. Cotton occupies a third of the area farmed, maize, millet and sorghum takes up half, while the rest is cultivated with groundnuts and other minor crops. Households still have some fallow land available, as well as access to communal woodland and grazing resources around the village. Since 1988, it has been supported by the farming systems research team at the Sikasso regional research centre, who have been collecting basic data to monitor socio-economic changes. They have also carried out on-farm trials on agricultural, livestock and forestry issues, including soil-fertility, grazing land management, as well as helping to establish a local land management agreement (Joldersma et al, 1996).

Rainfed cereal farming (research sites: Siguiné and Dilaba)

Outside the Office du Niger and CMDT areas, farmers grow bullrush millet (*Pennisetum typhoides*), sorghum (*Sorghum bicolor*), maize (*Zea mais*), cowpea (*Vigna unguiculata*) and minor crops like fonio (*Digitaria exilis*), dah (*Hibiscus cannabinus*) and groundnuts (*Arachis hypogaea*). In these drier areas, farmers have had to develop a diverse range of activities, including animal husbandry, migration, trading and crafts. Dryland farming systems rely heavily on livestock for provision of dung, and animal traction, as well as being a source of milk, meat, and capital. Two villages were chosen to represent rainfed cereal farming systems. Siguiné is located 30km south of Niono and still has considerable fallow land available. Dilaba, which lies 40km south of Segou, has much less land and therefore must develop other, often more labour intensive, strategies for maintaining soil-fertility.

The farming systems of Siguiné and Dilaba are dominated by the cultivation of millet for home consumption and sale of surplus. Dilaba benefits from higher rainfall than Siguiné but, with its higher population density, has less available land. Siguiné still maintains 'in-fields' or 'village fields' which are small permanently cultivated plots on loamy soils around the village which receive some manure and other organic matter, and where millet is intercropped with cowpea. 'Out-fields' or 'bushfields' are often several kilometres from the village on sandy soils cultivated without any nutrient inputs, and fallowed after 4 or 5 years' use. Land is sufficiently abundant for bush fields to be left uncultivated for up to 15 years, farmers with larger landholdings having a higher proportion of their bushfield land under fallow than other households.

In Dilaba, the millet/cowpea crop is usually grown in rotation with groundnuts or Bambara groundnuts (*Voandzeia subterranea*), leguminous crops being of greater importance in Dilaba than the other research sites. Groundnuts used to be an important cash crop but now cover no more than 5 per cent of field area. Fonio and sorghum are other minor crops. Fallows have all but disappeared and are now mainly used as space for oxen to graze. Dilaba faces particularly difficulties in maintaining a viable farming system since fallowing and nutrient transfers from grazing to cropland are now severely constrained. Sheep and goats are of particular importance to farm households as generators of manure and of income, and are better able to survive on the limited grazing available. But, in many years, cattle must spend several months of the dry season away from the village because of pasture shortages, thus reducing supplies of dung available for households to improve the fertility of their soils.

Due to dryland areas being considered of lower potential, they have not received support from bodies like the Office du Niger or the CMDT. In some places, a project or NGO has set up operations, from which villages can gain some temporary benefit, but overall, these areas have received much less attention and systematic assistance. Extension services outside the CMDT zone have been provided by the Programme National de Vulgarization Agricole (PNVA), established in 1990 by the World Bank, but for which funding ended in 1998. Using a training-and-visit approach, the PNVA organized the demonstration of new techniques but was not involved in credit or input supply. These dryland zones seem to have received more attention in colonial times. For example, one of the case study villages, Siguiné, was visited regularly by a French technician working for the colonial government. In those days Siguiné was still producing cash crops such as cotton and groundnuts, but this is no longer the case due to declining rainfall. The technician encouraged the villagers to start using organic fertilizer and introduced technologies such as compost pits and the use of litter in cattle *kraals*. The large-scale production of compost was, however, abandoned in the 1940s when the technician left the area and has only been taken up again in the last few years.

Dilaba, by contrast, received no technical advice during the colonial period. In the 1970s, Dilaba had access to the groundnut-promoting scheme Operation Arachide and Cultures Vivrières, a government agency which provided access to credit for inputs and equipment as well as marketing. Between 1986 and 1998, Dilaba gained support from the IFAD-funded Projet Fond de Développement Villageoise de Segou (PFDVS). But the credit programme targeted towards purchase of oxen and fertilizers was interrupted for several years, following farmers' failure to repay credit. PFDVS continued supporting other activities, such as sheep fattening, cereal banks, supply of fungicide, and tree planting until its funding ended in 1999. Dilaba also lies close to the agricultural research station at Cinzana from which short-cycle millet varieties have been obtained.

A COMPARISON OF THE FOUR RESEARCH SITES

The four village research sites therefore represent the main farming systems found in Mali (see Figure 3.1). Table 3.1 compares the main characteristics of the four villages, which provide contrasting locations, constraints and opportunities.

As can be seen from Table 3.1, population density is highest in Tissana, which benefits from the irrigation scheme, and lowest in Siguiné. Average landholding size is in inverse proportion to population density, decreasing from 44ha in Siguiné to 3ha in Tissana. Such differences in land availability

Table 3.1 *Socio-economic and physical characteristics of the four villages*

	Tissana	*M'Péresso*	*Dilaba*	*Siguiné*
Date of settlement	1955	around 1800	around 1880	around 1900
Average rainfall (mm)	350–550	750–850	650–720	350–420
Main soil types	Clay	Loamy sands and sand	Sand and clay	Sand and clay
Population (n)	1303	850	243	502
Population density (n/km^2)	50	18	32	6
Mean household size (n)	15	15	27	18
Mean landholding size (ha)	3	18	17	44
Main crops	Rice, vegetables	Cotton, maize, sorghum, millet	Millet, cowpea	Millet, cowpea
Mean cattle holding (n)	5.5	12.2	20.0	16.0
Mean holding sheep and goats (n)	0.5	12.0	34.0	15.0
Distance to market	12km to Niono (all-season dirt track)	25km to Koutiala (dirt track)	40km to Segou (mainly asphalt)	30km to Niono (mainly dirt track)

have major implications for how farmers manage soil-fertility and, in particular, the role played by crop–livestock interactions. Most households possess some cattle and small ruminants, but numbers vary considerably between households (see also Table 3.3). Cattle are important both as a store of value, and for the inputs they provide to the cropping system, particularly traction for ploughing, and dung for fertilizing the land. Throughout Mali, farmers have been investing in cattle so that nowadays, a major share of the national herd is owned by sedentary farmers. However, the animals are often put in the care of hired herders, so that farmers can concentrate on farming. Tissana and Siguiné farmers rely on hired Fulani herders to care for cattle other than those needed for ploughing and milk. However, in Dilaba and M'Péresso, farmers themselves guard their animals.

Households in the four research sites are also engaged in a range of off-farm activities, which provide an important source of income, particularly for individuals. For villages relatively close to a small town, this provides opportunities for petty trade and other sources of employment, such as production of crafts for sale and work for others (transport, brick-making, wage labour, etc). In addition, migration has been, and continues to be, a major source of cash by which means households can pay tax, purchase new equipment and oxen, and provide for wedding expenses. It is particularly important in Dilaba and Siguiné, where the young men often go for several years to Ivory Coast or Bamako. Household heads in both villages complain about the difficulties they face in collecting and transporting enough organic fertilizer during the dry season, since most of their young male members are away earning cash on migration. By contrast, young men in villages like Tissana in the Office du Niger often spend the dry season at home, since irrigated vegetable production now provides a valuable source of additional income.

Soils and their management

Soils in Dilaba and Siguiné are similar and comprise mainly wind-deposited sandy soils, with smaller areas of clay soils developed from sedimentation. M'Péresso's soils are loamy clays in lower lying areas and sandy soils on slopes and uplands. The irrigated fields of Tissana are all situated on medium and fine clays. In all sites, soils have low pH levels and a relatively low organic-matter content, as with many soils throughout the Sahel (Table 3.2). Nitrogen and phosphorous levels vary from low to very low. Tissana, with its clay, has the least problematic conditions. Potassium levels are particularly low for M'Péresso, low for Tissana and Dilaba, but reasonably good for Siguiné. Infiltration capacity is generally good for all the villages, except for Tissana, while erosion risks from rain and wind are significant for Dilaba, Siguiné and M'Péresso (Dembélé et al, 1998).

Farmers' classification of soils and perception of changes in soil-fertility

The landscape around the village is usually divided into different areas according to characteristics of the soils and major differences in relief. Minianka farmers around Koutiala, where M'Péresso is situated, distinguish between soils according to colour, texture and ease of cultivation (Kanté and Defoer, 1994). The uplands are characterized by thin, gravelly soils with many coarse elements (*niangua*) and are generally used as communal pasture. Down the slope, *niang-tiôon* soils are found, which are sandy gravels, which used to be preferred because they were easy to prepare. With the introduction of the oxen-drawn plough, *niang-tiôon* soils were left to woodland and pasture, although due to land pressure, these soils are being cultivated again. Sandy soils (*guechien*) are favoured because they are easy to cultivate, but they are

Table 3.2 *Physical and chemical characteristics of the soils found in the four villages*

	Tissana		*M'Péresso*		*Dilaba*		*Siguiné*	
Soil structure	Clay		Loamy sands and sand		Sand and clay		Sand and clay	
Soil types (FAO)	Gleysols		Gleyic acrisols, eutric gleysols, haplic arenosols		Haplic, arenosols, gleysols		Haplic, arenosols, gleysols	
Soil depth (cm)	*0–20*	*20–40*	*0–20*	*20–40*	*0–20*	*20–40*	*0–20*	*20–40*
N per cent	0.02	0.01	0.01	0.01	0.01	0.01	0.01	0.01
P/ppm	6.00	4.00	3.20	0.90	2.20	1.40	1.70	1.20
K	0.23	0.28	0.12	0.12	0.23	0.18	0.36	0.24
pH (KCl)	5.00	5.20	4.20	4.10	5.10	4.60	4.70	4.90
OM per cent	0.49	0.22	0.22	0.18	0.09	0.14	0.15	0.11
CEC % meq	11.70	16.30	2.50	4.30	2.60	3.80	4.60	6.40

Source: Dembélé et al, 1998
Key: pH = acidity/alkalinity; OM = Organic Matter; CEC = Cation Exchange Capacity

easily exhausted when not regularly fertilized. With the general introduction of the ox-plough, the heavier and more fertile *tawogo* soils are now possible to cultivate. The lowest part of the catena, known as *fâa*, is made up of heavier clays, with some waterlogging and flooding, where farmers plant trees and vegetables.

With the lower rainfall of recent decades, farmers start by cultivating the sandy soils which can be worked with the first showers of the season. By contrast, heavier clays can only be ploughed once they have become much wetter, later in the season. The upland gravels are particularly vulnerable to erosion and require soil conservation measures to ensure their continued productivity.

Farmers in M'Péresso have set aside certain parts of the village terrain (mainly *niangua* and *niang-tôon*) for pasture and woodland, while concentrating cultivation in areas with the best soils (*guechien* and *tawogo*). The farming areas most prized today are not necessarily those most highly considered in earlier times. For example, before the 1973–1974 drought, the heavier, more fertile *tawogo* soil type was not used for farming because of waterlogging during the rainy season. However, it is now highly prized for growing cotton and cereals.

The same criteria for distinguishing different soil types were found in the other study villages. In Dilaba (Figure 3.2), for example, farmers talk of *cencen* sands, found mainly to the south-west and east of the village territory, used principally for millet and groundnuts. These soils were the first to be farmed because they were easy to cultivate with hand tools. Once ploughs became available in the 1950s, farmers started using the clays and loams of the *ja* (hard soils). The black clays (or *bwafin*) are particularly suited to sorghum because of their better water-holding capacity.

Farmers use various indicators, such as colour, to assess soil-fertility. A black soil is considered fertile while a change in colour towards red or white indicates diminishing soil-fertility. Other indicators used include a decline in yields, changes in water holding capacity, and the emergence of certain plants, such as *Calotropis procera* (*fogofogo*) and *Striga*.

Farmers in all four villages consider the droughts of the 1970s to have been the turning point for a range of environmental factors and levels of productivity. Erosion is also seen as another cause of falling productivity, while lack of availability and high cost of mineral fertilizer are identified by farmers as obstacles to improving this situation. In Siguiné, farmers point to the early 1970s as a period of great difficulty when the harvest failed three years running. Tree cover has become much thinner since then, and wood for cooking and construction is increasingly scarce. Formerly, the bush provided a wide variety of fruits, leaves and nuts for home consumption and sale, but nowadays women say they can gain much less of value when they go gathering (Dembélé et al, 1998). Thus, villagers are convinced that a major part of the difficulties they face is due to the decline in rainfall levels, a perception also clearly expressed by farmers elsewhere in Mali (Brock and Coulibaly, 1999; Ramisch, 1999; Toulmin, 1992). At the same time, they acknowledge that their difficulties have been aggravated by the expansion of fields into former grazing lands, so that the village territory is gradually filling up.

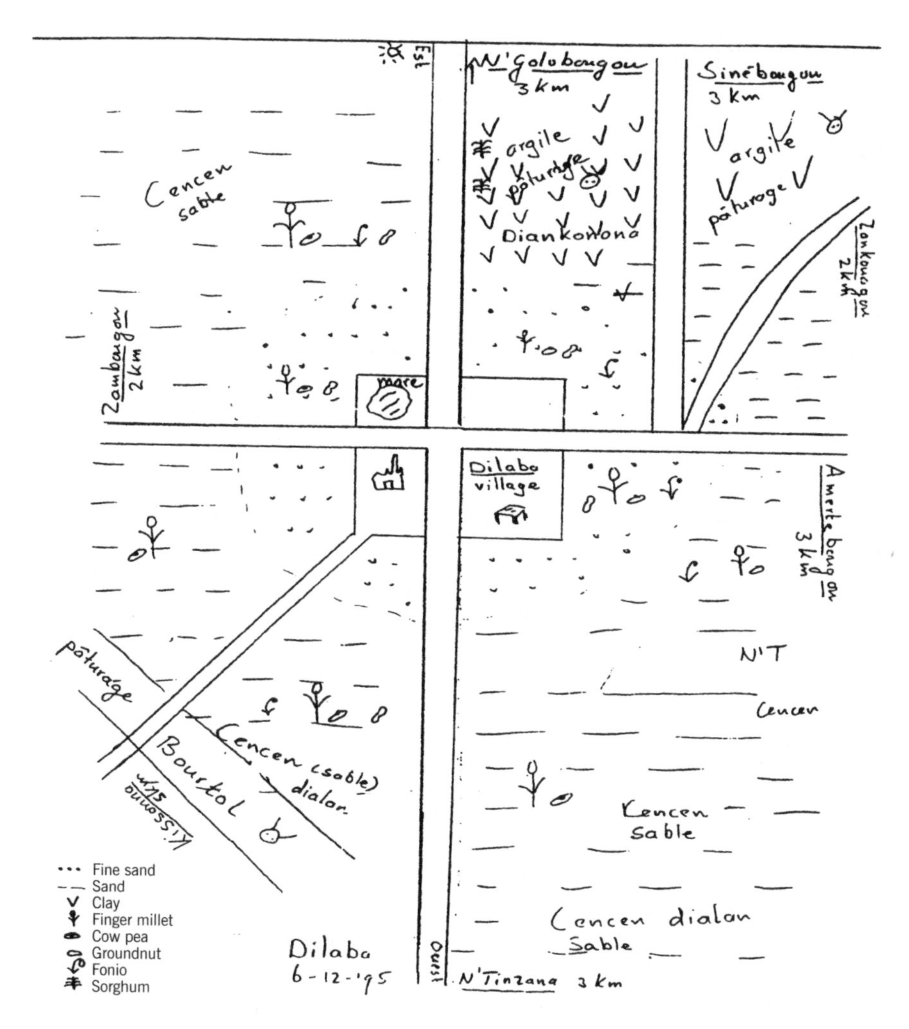

Figure 3.2 *Village territory map, Dilaba*

Differences in soil-fertility management strategies

In order to identify key characteristics associated by farmers with good soil-fertility management, people in each of the sites were asked to specify what, for them, constituted good practice. These were:

- Application of organic manure;
- Re-use of crop residues (mainly as bedding or as fodder);
- Use of mineral fertilizer;
- Anti-erosion measures (particularly in the cotton zone); and
- Maintenance of canals and ditches (in the irrigated rice zone).

Farmers identified four factors which help explain differences between households regarding how effectively such practices are carried out. These are the

Table 3.3 *Socio-economic diversity between farmers in the four villages, by resource group*

	Tissana			*M'Péresso*			*Dilaba*			*Siguiné*		
Resource group	*I*	*II*	*III*	*I*	*II*	*III*	*I*	*II*	*III*	*I*	*II*	*III*
N	*8*	*8*	*4*	*3*	*9*	*7*	*3*	*2*	*5*	*4*	*4*	*4*
Family labour/ household	17.0	8.0	6.0	13.0	7.2	4.4	29.3	8.5	2.4	24.0	11.0	6.0
Landholding (ha)	6.2	2.7	1.8	28.0	20.4	10.0	31.5	18.0	8.0	80.0	31.0	20.0
No of oxen	4.7	2.4	0.5	6.0	3.4	2.6	14.0	4.0	1.2	9.3	4.0	1.8
No of cattle (oxen included)	8.0	4.0	0.7	27.0	12.7	5.1	73.0	4.0	1.4	32.0	6.6	2.5
No of sheep and goats	0.5	0.7	0.1	24.0	11.2	7.6	138.0	35.0	6.0	30.0	14.7	7.7
No of carts	1.2	1.1	0.4	1.7	1.0	0.9	2.0	1.0	0.6	1.5	1.3	1.0

number of animals owned, possession of a cart, availability of labour, and the general level of motivation and energy within the household to invest a lot of time in making and transporting organic manure. Using these characteristics, households in each site were classified by villagers into three main groups. These resource groups have been used as the basis for the subsequent data collection and analysis of soil nutrient flows.

- Resource group I – those who manage soil-fertility well.
- Resource group II – those who are average managers of soil-fertility.
- Resource group III – those who manage soil-fertility poorly.

This classification of households demonstrates the wide differences found within each village in terms of access to key resources, such as labour, land and livestock. Households belonging to resource group I tend to have much more family labour available and cultivate a larger area. Their herds are also considerably larger than for the other two resource groups. Poor soil-fertility managers, as might be expected, are the least well-endowed farmers and form about 40 per cent of all households in the sample. Particularly in Dilaba, there is a large group of fairly small poor households with access to only limited means of production. These farmers are less involved in growing millet, and are concentrating their resources on a broader range of crops including fonio, sorghum and cowpea (Dembélé et al, 1998). The contrasts between farm households can be seen from the cases presented in Box 3.1, which demonstrate the wide differences in assets, opportunities and strategies between farmers.

How do people gain access to soil nutrients?

Soil-fertility management practices include the use of mineral fertilizers, the production of organic fertilizers from cattle pens, waste heaps, composting and cattle kraaled overnight on harvested fields, as well as erosion control

BOX 3.1 CASE STUDIES

Farmer A (resource group I, Dilaba) is recognized as one of the best in the village of Dilaba. He has 60–80 cattle, including 12 oxen, about 30 sheep and goats, and enough ploughs and carts to cultivate his 22 hectares of millet. Although there are 30 people in his household, he does not have enough family labour to cultivate his fields. His first children were all girls who are now married and living elsewhere. His sons are still young so he must hire 5 labourers to help with the farming. From his 22 hectares, he harvested 30 tons of millet, 500kg of sorghum, some cowpea, and a bit of sorghum intercropped with the millet. His cattle herd generates a lot of manure. During the dry season, his herd pass the night in a cattle *kraal* on one of his fields, while the oxen are kept in a pen close to the homestead where they receive supplementary feeding with crop residues so that they will be in good form when the ploughing starts. He produces an increasing amount of organic manure, made up of cow dung and household waste averaging 50 tons in the period studied. He also has several waste heaps to which he adds all the dung from the sheep and goats. In addition he hires labourers to collect weeds to add to the waste heaps. He cuts the branches and leaves of small bushes growing in his fields and works these into the soil. He used to burn the residue left on his fields before ploughing but has now stopped, following discussions on how to maintain soil-fertility with the research team. He no longer uses mineral fertilizer, due to the uncertain rainfall and risks of crop burning.

Farmer B (resource group III, M'Péresso) had no agricultural equipment nor oxen until 1997, when his sons' migration earnings permitted the purchase of a cart and a pair of oxen. Although he is old, he has not had many sons, which has limited his farming activities. He has a few head of cattle, 5 sheep and goats and 7.5 hectares of land, all of which is cultivated. Total production in 1996 comprised 3.3 tons of cotton, 4.6 tons of cereals, and 1000kg of groundnuts. He uses chemical fertilizer for his cotton, but below the recommended rate because of its high cost and because part is used on the maize crop. He has a waste heap, like everybody else in M'Péresso, and also produces manure from his cattle and small ruminants. His organic matter production increased from 3 to 6 tons over the course of the research. The sale of honey provides a significant supplementary source of income.

Farmer C (resource group III) came back to the village of Siguiné about 10 years ago. His small household stems from a larger extended family which broke up. He only has in-field land close to the village, 13ha of which is under millet and 5ha in fallow. He is poor, and has only two oxen, one plough and two carts, while his wife has half a dozen sheep and goats. Only one child is old enough to help him farm. He works hard as a blacksmith, which enables him to supplement his meagre harvest. His yields are low, at 150–350kg/ha, depending on rainfall levels. With only a small area of fallow, Farmer C cannot rely on fallowing as a means to restore fertility to the soil. He recognizes the benefits of cowpeas grown in association with millet as a way of stemming a rapid fall in soil-fertility. His own production of organic matter was low due to the limited number of animals owned and averaged 100kg/ha. He depends on household waste and droppings from his sheep and goats, mixed with grasses cut in the bush.

Farmer D is a well-off farmer (resource group I) from Tissana. He farms 7.4ha of land, of which 2ha is double cropped, as well as owning more than 50 cattle which have been put into the care of a Fulani herder. Six plough oxen are kept at his homestead, as well as several sheep and goats. He has so much land because he was one of the first to settle in Tissana, and had many family workers at the moment when the Office du Niger was allocating land. He also has some land on the edge of the irrigation scheme used for vegetable cultivation (onions, tomatoes and sweet potato). His wives and sons have vegetable plots of their own, which produce an important source of dry season income. Farmer D harvested an average of 32 tons of rice, giving a yield of 4.4t/ha on the single-cropped land, a figure broadly comparable to mean yields across the Office. Mineral fertilizers are applied in line with the recommended rates (200kg/ha urea and 100kg/ha DAP). The stall-fed oxen are fed with rice bran, and the manure is stored for use on the vegetable plots. During the rainy season, the cattle herd is far away, though they come back once the rice is harvested. He has constructed a hangar to stock crop residues with help from the research team, designed to avoid termite attacks, which have posed a very serious problem in the past. Although advised to stop burning the stubble and compost this instead, he wonders how else to clean the field prior to ploughing. He regards this as too labour-intensive since it competes with vegetable production.

Table 3.4 *Use of various soil-fertility measures by households in the four villages*

	Tissana	*M'Péresso*	*Dilaba*	*Siguiné*
Mineral fertilizer	100% (rice)	100% (cotton and maize)	very limited	very limited
Manure	40–80%	100%	100%	100%
Compost	limited	30–60%	very limited*	just started*
Household waste	yes	yes	yes	yes*

Note: * Villages started as a result of the participatory research and on-farm trials conducted by ESPGRN

measures and maintenance of bunds in rice fields. Table 3.4 gives an overview of the popularity of these different practices in the four villages.

Mineral fertilizers

Mineral fertilizers have been used by all cotton and rice farmers since the 1950s. The CMDT and the Office du Niger propagate a recommended rate for fertilizer application which is generally known by farmers. The CMDT also has a recommendation for levels of organic fertilizer use which takes into account the fact that this tends only to be applied at the beginning of the rotation with cereals grown the year after. Cotton farmers use urea (N46) and a complex fertilizer (N:P:K:S:B 14-22-12-7-1),[8] adjusting the amount applied depending on the fertility of the plot and the soil type. They also adjust the urea dose if the crop is sown late. Farmers who use a lot of organic fertilizer are likely to use less than the recommended rate, as can be seen from resource group I farmers in M'Péresso who use a lot of organic fertilizer, but only apply 60 per cent of the recommended rate for the cotton formula fertilizer.

Rice farmers use urea and DAP (N:P 8-46). Farmers know about the difference in effect of the various types. Rice farmers, for example, know that DAP stimulates tillering of rice, and urea stimulates the filling of the grains. Farmers have a less clear view on the difference between the nutritive elements N, P and K, and are mainly familiar with potash.[9]

Rock phosphate is produced by mines in the Tilemsi valley, in the north-east of Mali. However, at present, it is hardly used by farmers, although campaigns promoting its use were carried out in the 1990s. Currently, plans are afoot to transform these rock phosphate supplies into bagged fertilizer through the construction of a blending plant at Markala on the river Niger, close to Segou (Gakou et al, 1997).

Dilaba farmers started using mineral fertilizers in the 1970s for groundnuts when first the OACV and thereafter the PFDVS provided inputs on credit, as described earlier. Since the PFDVS stopped supplying fertilizer on credit in 1987, the villagers have hardly used any mineral fertilizer because of the low response and unfavourable price ratio of fertilizers in relation to rainfed cereals. Moreover, the low and unpredictable rainfall makes its use risky. Mineral fertilizers have never been widely used in Siguiné. Thus, today, only a few farmers in Dilaba and Siguiné use mineral fertilizers, and in extremely small quantities, for example for *Striga* control or mixing with seed.

Manure

Manure has been used by all farmers in Siguiné, Dilaba and M'Péresso for many years. Formerly, it was barely used in the rice village, Tissana. Until recently, organic fertilizers were not a priority recommendation for extension workers in the Office du Niger, but this has changed, following the rapid increase in fertilizer prices after the devaluation of the CFA franc in 1994. Farmers have become increasingly interested in using organic fertilizer on rice, but the demand for organic fertilizer is particularly linked to the expansion of dry season vegetable growing. As a result, the subsequent rice crop benefits from the residual effect of the manure applied.

Tissana farmers rely on Fulani herders to care for a large part of their herd, their cattle staying some 10–20km outside the irrigation scheme during the rains and only coming to graze the rice stubble during the dry season after harvest. Oxen and a few milking cows are kept around the village, being stall-fed or herded with great care to avoid risk of crop damage. Some farmers also arrange with their hired herder to bring their cattle into the scheme for the night, to increase access to manure then used for the cultivation of vegetables. In addition, farmers go and fetch manure from the *kraals* of Fulani herders on the edge of the irrigation scheme, though a small payment is usually required if their animals are not in the care of that herder.

The herds of farmers in Siguiné and M'Péresso normally stay the whole year in the village. Resource group I farmers in Siguiné fertilize their land mainly by leaving their cattle overnight in selected in-fields. This method is also used in the other villages when the herd is large enough. In Dilaba, cattle pass the night on the fields for 3–4 months, but once fodder becomes scarce, the cattle must go elsewhere to graze. Where kraals are established on the field itself, this method saves on labour investment in transport of the organic matter.

In contrast to many other dryland villages in Mali, in none of the research sites were farmers involved in manuring contracts with transhumant herd-owners (Toulmin, 1992). Under such contracts, herders will keep their livestock overnight on a selected field, in exchange for access to stubble, water and occasionally, provision of food for the herder. Even farmers without many animals of their own can gain appreciable amounts of manure from this system. Villagers recognized that, although these contracts had been more important in the past, farmers in Dilaba do not have enough land available to offer grazing to others. Reasons for the absence of such contracts in Siguiné include the relative scarcity of water, and heavy pressure on grazing from village and neighbouring herds. In addition, transhumant herders prefer the option of spending time on the rice stubble of the Office du Niger. In the case of M'Péresso, villagers prefer to keep village grazing lands for their own animals, given shortages of fodder.

Manure is also produced in cattle pens, situated either at the homestead or near the field. Farmers in M'Péresso construct pens for both the dry and wet season, the latter having been introduced by the Sikasso research team and are mainly confined to farmers who participated in the joint on-farm trials. In Tissana there are only dry season cattle pens. Production increases when bedding material, often crop residues, are added to the pen, a practice which has now become widespread in M'Péresso, and helps account for the substantial quantities of organic fertilizer they produce (see Table 3.5). Recycling of crop residues through these cattle pens is a recommendation made by the extension service and is widespread amongst farmers in the cotton zone. In Tissana, work oxen and donkeys are stall-fed for much of the year, and their manure mixed with rice straw helps maintain fertility of rice fields as well as returning potassium to the field. However, the quantities produced remain small.

Composting of manure is rarely done in any of the villages studied, despite the fact that poorly decomposed manure is likely to reduce the availability of N through N-immobilization. Farmers recognize the risk from animal wastes of crops being 'burned', due to high concentrations of nitrogen from urine and dung, particularly when have remained for a long time on the same patch of field. The risk of burning is particularly a problem in drier areas such as Siguiné and Dilaba. Where animals are not kraaled directly on the field, manure is transported to the fields in the dry season, and deposited in heaps to be spread out before ploughing. However, this can be a time-consuming activity for which a cart is essential. Manure produced by sheep and goats is used most extensively in Dilaba where small stock are particularly numerous. Their dung is considered more effective than cattle manure because of its composition and form, but takes longer to release its nutrients as the droppings are dry and hard to break down. Given the shortage of manure, some farmers in Dilaba now pay children to go and collect dung deposited by cattle in the common grazing areas to be brought back and spread on their fields.

Composting, recycling, household waste

All farmers in the Office du Niger incorporate any remaining crop residues at the start of the farming season, although a large proportion of these will have been consumed by livestock by then. In M'Péresso they are either burned for making potash, or used as litter in cattle pens. Compost-making is a technique which has recently been re-introduced and widely taken up within the CMDT area. Farmers in M'Péresso are better able to make compost than those in the other three villages studied because it benefits from higher rainfall and easier access to water.

In Siguiné, crop residues are grazed and used for making potash, with the remaining stubble being burned at the end of the dry season in preparation for sowing. In Dilaba, crop residues are harvested and recycled with more care, given the scarcity of natural grazing. As a result, burning of the residue is infrequent. While farmers acknowledge the potential benefits from composting organic materials, they face serious constraints to producing and transporting organic fertilizer, due to limited availability of labour, livestock, transport, and water. Labour is a particular constraint on production of organic material for smaller households which are mostly found in resource group III. The transport of crop residues to the cattle pen or compost heap, and of manure to the fields, mainly takes place in the dry season when young men are away on migration. Few household heads are prepared or able to pay those who stay behind to produce and transport organic fertilizers in the dry season. Small, poor families face major difficulties in trying both to maintain a productive farm enterprise with its multiple components and associated activities, and to benefit from cash earnings through dry-season migration.

In the Office du Niger, the dry season has now become a busy period as all family members have become involved in the cultivation of vegetables, often for their own benefit, rather than as a contribution to the income of the larger household. They use manure from smallstock and household waste on these vegetable plots, particularly for onions and tomatoes. As a result, the household head may face a shortage of available labour who can be called on to produce and transport organic materials to the rice fields.

Access to transport is important for making and spreading organic fertilizer. Farmers rely on donkey-drawn carts, although some of the poorest resource group III farmers in M'Péresso had to carry material in baskets to the fields. Carts have become a valued farm asset and are also used for transporting harvests, water and bricks for house building, as well as going to market. However, while not all households possess a donkey cart, it is often possible to rent or borrow this equipment, although this may not be available at the preferred time.

Table 3.5 below presents the amount of organic fertilizer produced per hectare for each of the four villages and by farmer resource group, which demonstrates clear contrasts between farmers and location. As can be seen, manure use in Tissana is still very limited, whereas farmers in M'Péresso have taken up this soil-fertility measure and are using it intensively. Richer farmers with larger livestock holdings are able to generate substantial quantities but

Table 3.5 *Organic fertilizer production and use (kg/ha) for the four villages and by farmer resource group. Average over two farming seasons, 1995–1997*

Village	*Resource group I*	*Resource group II*	*Resource group III*
Tissana: Rice	173	250	282
M'Péresso: Cotton	3576	1950	1755
Dilaba: Millet	1727	448	644
Siguiné: Millet	623	546	412

Note: Data for Siguiné refers to a single farming season 1996–1997, and refers only to the in-fields around the village. Outfields receive no fertilization

even resource group III farmers are producing as much per hectare as the highest producers in Dilaba. In the millet-producing villages of Dilaba and Siguiné, organic fertilizer provides by far the most important source of soil nutrients, but most farmers have relatively limited access. The similar levels found for all farmer resource groups in Siguiné hide a difference in source, with resource group I farmers relying almost entirely on cattle herds, while resource groups II and III must resort to spreading household waste on their land. Resource group I farmers in Dilaba comprise several households with very substantial holdings of cattle, sheep and goats, which provide large quantities of dung for their land.

Most farmers cannot produce sufficient organic fertilizer to meet all their needs and therefore have to set priorities according to the fertility of a field, soil type, crop to be grown, location of the field, and its tenure status. In M'Péresso, most organic fertilizer is applied to cotton, which is grown in rotation with cereals. Farmers combine mineral fertilizer with organic manure to increase its quality and to prevent acidification of soils. Preference is given to the least fertile parts of the field and to sandy soils which are subject to water erosion and considered to promote the best response. Maize and other cereals benefit given that they follow cotton in the rotation. In Siguiné, organic fertilizer is only applied to the in-fields which are under permanent cultivation. Dilaba farmers give priority to less fertile spots where yields are low and the soil colour has lightened, which is considered a sign of decreasing soil-fertility. In Tissana, organic fertilizer is applied only to those parts of the rice field where vegetables are grown in the dry season.

The tenure status of a field might be expected to affect levels of fertilizer application. Where a farmer has land on loan from a neighbour, he may be discouraged from investing much effort in adding organic matter to the field, since such inputs have a tendency to provide residual benefits over a number of years. However, no differences were found in levels of fertilizer use whether mineral or organic, on hired or owned land, providing evidence that tenure does not seem of major importance for soil-fertility management in any of the four villages.

MANAGING SOIL-FERTILITY: ANALYSIS OF NUTRIENT BALANCES

An analysis of soil-nutrient balances was carried out for each village to assess patterns of nutrient use and distribution across different fields, crops, households and locations as a means to understand farmer decision making. It was undertaken by working with a selection of farmers from each resource group for all four sites. Farmers began by drawing resource flow maps of their overall holding, its main components and resource flows. This map then provided the basis for measuring and monitoring the flow of inputs and outputs from the farm holding, as well as internal flows between components of the farm. During subsequent cropping seasons, farmers drew up planning maps, which were then amended by indicating which activities had been carried out (Defoer and Budelman, 2000).

Information from these maps, drawn over three consecutive seasons, was also entered into the computer, the data forming the basis for analysing nutrient flows and calculating nutrient balances. A simplified version of the nutrient-balance model developed by Smaling (1993) was used (Defoer et al, 1998), allowing the calculation of partial nutrient balances. The crop production system was taken as the unit of analysis, with the following functions taken into account.

- IN1 = mineral fertilizer applied to crops.
- IN2 = organic fertilizer (manure, household waste, compost etc) applied to crops.
- OUT1 = harvested grain.
- OUT2 = crop residues removed from the fields.

Partial nutrient balance = IN1 + IN2 – OUT1 – OUT2

Measurements of inputs and outputs were mainly derived from the farm maps, combined with additional information obtained from sampling. The data were collected over two farming seasons, using local units and then converted into kg.[10] The nutrient content of the different products were calculated using conversion ratios taken from the literature. Such partial nutrient balances need to be interpreted with care (Scoones and Toulmin, 1998), since inevitably there will be measurement errors and estimations at each stage in the calculations (see Chapter 1). However, these partial nutrient balances, in conjunction with resource-flow maps at farm level, can provide a very useful means for understanding how the farming system works, and indicating areas for improving nutrient-use efficiency.

Here, the nutrient-balance data from the cash crop and dryland cereal systems will be discussed to help identify significant differences between crops, fields, and farmer resource groups. As will be seen below, there is a marked distinction between farmers producing cash crops (cotton, rice) in the CMDT and irrigated areas, where access to mineral fertilizer is relatively easy and crop sales provide sufficient means and incentive to invest in mineral fertilizer. By contrast, farmers in rainfed cereal growing areas exhibit nutrient balances which are consistently negative, given their reliance on limited amounts of organic matter. However, in each of the villages, there is considerable diversity between households according to their access to resources and consequent management practices, although the diversity between sites seems to be much greater than that found within a single location.

Cash-crop producing areas: rice and cotton

Figure 3.3 presents a resource-flow map for a farmer from resource group I in the cotton-growing village of M'Péresso. It shows a farm-holding of 20ha, of which 4ha are under cotton, 10ha under cereals, and the remainder under groundnuts and fallow. A total of 17 sacks of fertilizer have been applied to a range of crops, principally cotton, but also sorghum, maize and dolichos bean. As with other resource group I farmers in M'Péresso, he generates substantial quantities of organic fertilizer, as was seen in Table 3.5, through an extensive web of crop residue recycling activities. Components of this system include grazing of stubble, storage of residues for subsequent use as fodder by cattle and donkeys, and establishment of compost pits. Ownership of at least one donkey and a cart is essential to support such an intensive movement of organic materials around the farm. Despite such activities, when the combined cotton-millet rotation is considered, while N is in slight surplus, there are outflows for P and, especially, K.

Table 3.6 below provides average figures for N, P and K flows for the two villages of M'Péresso and Tissana. As can be seen, the partial nutrient balances are positive for all farmers and nutrients on land under cotton in M'Péresso. Cotton receives the major share of nutrient inputs in M'Péresso, with organic matter of equal importance to mineral fertilizers for richer farmers, so far as nutrients are concerned, though of lesser importance for poorer households. With average cotton yields of 1100kg/ha (ranging from 400 to 1800 kg/ha), the nutrient balance for cotton was estimated to be positive in both years, for all resource groups of farmer and all nutrients. However, since cotton is followed in rotation by millet or sorghum, for which the partial nutrient balance is significantly negative for all household resource groups, the overall situation at the end of two farming seasons is much less promising. Sorghum and millet are grown with almost no extra inputs, save for a light application

Table 3.6 *Summary table of partial nutrient balances (kg/ha) for Tissana and M'Péresso, by farmer resource group, 1996–1997*

Village	*Resource group I*			*Resource group II*			*Resource group III*		
	N	*P*	*K*	*N*	*P*	*K*	*N*	*P*	*K*
Tissana: Rice	24	6	–91	38	9	–88	40	8	–86
M'Péresso: Cotton	41	2	12	33	3	1	30	2	2
M'Péresso: Millet	–35	–5	–58	–30	–4	–46	–19	–3	–31
M'Péresso: Millet and cotton	6	–3	–46	–3	–1	–45	10	–1	–29

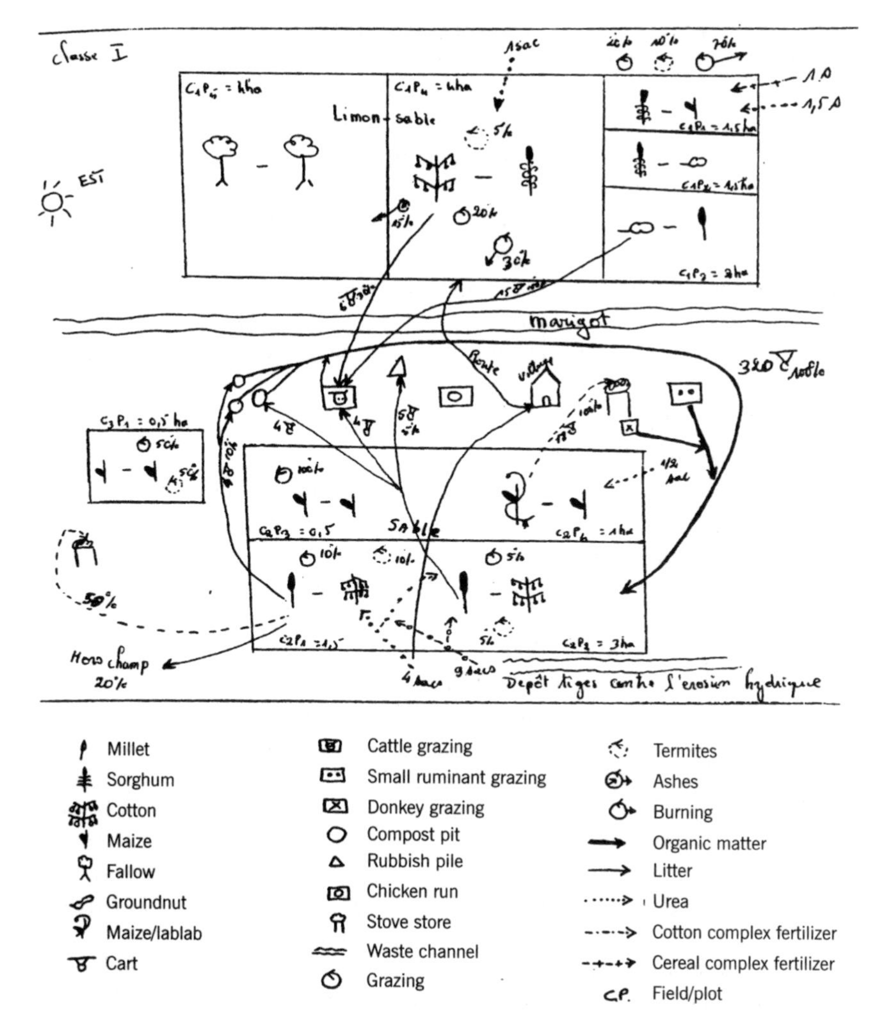

Figure 3.3 *Resource-flow map: resource group I farmer from M'Péresso*

of cereal-complex fertilizer by one household, and a few cartloads of organic matter by a couple of other households in resource groups II and III. Yields of sorghum and millet averaged 1000kg/ha and 700kg/ha respectively for the two years studied, which account for a major share of the nutrient outflow, with other outflows stemming from grazing of stubble, collection of straw for fodder, burning of stubble for making potash, and use in compost pits. Losses of K from the system are particularly marked.

In the case of Tissana, the nutrient balance study focused exclusively on single-cropped rice plots. Levels of mineral fertilizer use are high for all farmers and often in excess of recommended rates. Crop yields show no significant differences between farmer resource groups, and averaged 4.5t/ha over the three-year period 1995–1997. As a result, partial nutrient balances are broadly similar, with a surplus of N and P accumulating on rice plots. By contrast, outflows of K are very substantial, the large part of which is the result of animals grazing on crop residues. Only a small share of crop residue is carried from the field as fodder for stall-fed animals, and only a small proportion of this outflow was returned to the rice plot through organic matter applications. Current stocks may be sufficient to support such outflows for a few years, but some means of addressing such a regular loss must be considered in the longer term. Analysis of vegetable production in Tissana did not form part of the detailed nutrient balance study. However, their importance has been growing very significantly, and it is likely that the subsequent rice harvest benefited from these organic inputs. Vegetable fields rotate their location from year to year, with farmers selecting less fertile spots for their cultivation.

Rainfed cereal villages

Figure 3.4 presents a resource-flow map for a resource group I household in Siguiné, to demonstrate the significance and direction of different nutrient flows. The farm holding is dominated by millet and cowpea, with only 1 per cent of land under groundnuts. The outfields of 28ha receive no fertilizer of any sort, although grazing of stubble by cattle during the dry season may contribute a little dung to the field as they move across the field. The in-field holdings total 21ha and receive whatever nutrients are available from the dry-season stabling of cattle on the field and the transport of manure from livestock kept around the homestead. Cowpea hay is stored for subsequent use, or sale as fodder. On many plots, a major share of the millet residue is burnt for potash. The farmer purchases no mineral fertilizer. As can be seen from Table 3.7, this farmer, as with all households in Dilaba, is unable to generate sufficient organic materials to restore soil nutrients to the land.

Table 3.7 presents the partial nutrient balances calculated for Dilaba and Siguiné. Resource group I farmers have more cattle, greater access to household labour and far larger areas of land under cultivation. However, as was seen in Table 3.5, levels of organic matter production per hectare of in-field were similar for all household resource groups in Siguiné. As a result, it is not surprising to find that grain yields for the in-fields of all farm households were broadly similar for 1996, the single year for which data was collected, at around 550–600kg/ha for millet. Average bush field yields were similar at 600kg/ha, but differed considerably between fields, due to variations in soil, number of years under cultivation, timeliness of sowing, and availability of labour and plough teams to achieve an early weeding.

Table 3.7 *Summary table of partial nutrient balances (kg/ha) for Dilaba and Siguiné, by farmer resource group*

System and year of observation	*Resource group I*			*Resource group II*			*Resource group III*		
	N	*P*	*K*	*N*	*P*	*K*	*N*	*P*	*K*
Dilaba, all farmland	–23	– 3	–33	–26	– 3	–31	–24	– 3	–34
Siguiné, in-fields	–18	– 3	–29	–16	– 2	–26	–14	– 2	–22
Siguiné, outfields	–40	– 6	–55	–24	– 3	–32	–31	– 4	–40

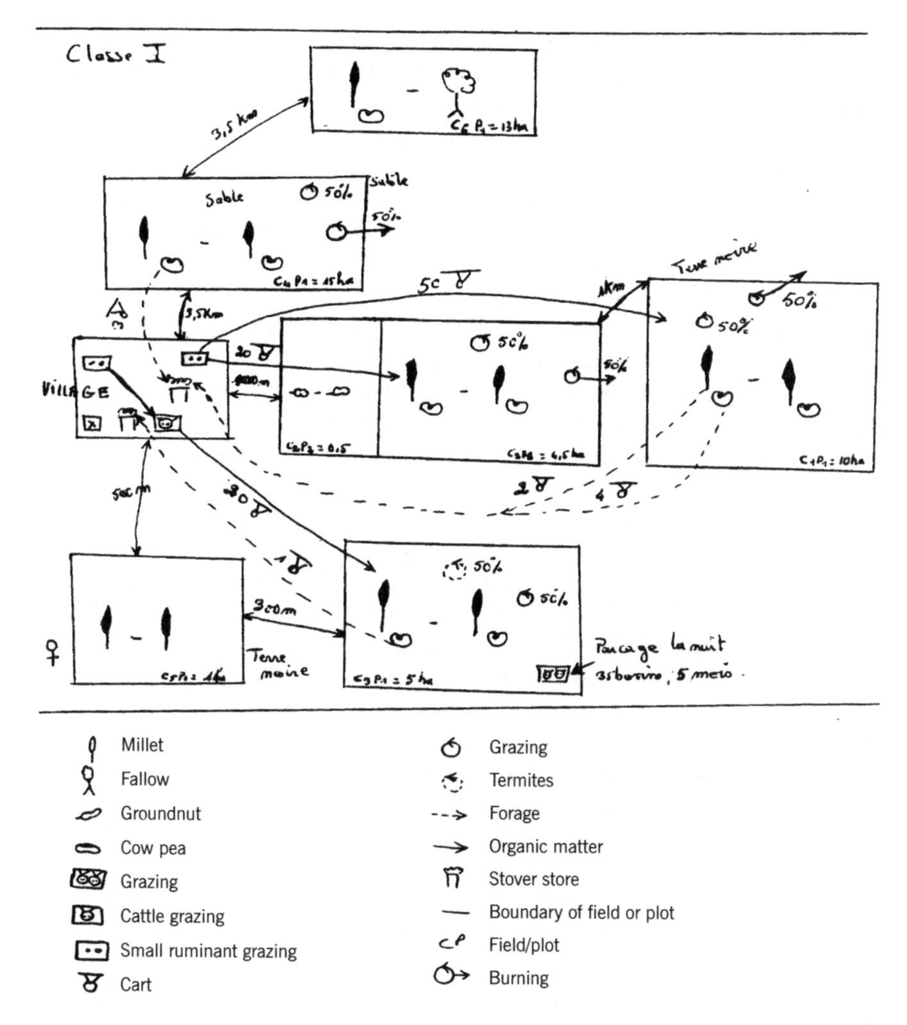

Figure 3.4 *Resource-flow map: resource group I farmer from Siguiné*

For both in-fields and outfields, losses of soil nutrients (N, P and K) from grain harvests represent only half of all such outflows, with significant amounts being taken off by grazing and making potash. Compost-making was insignificant in 1996, although this had increased significantly by 1997. For outfields, nutrient losses were more marked for resource group I households, a probable consequence of the higher yields they obtain because they can fallow land more often.

Dilaba's farming system used to follow a similar pattern to that of Siguiné, with a distinction between in-fields and outfields, and reliance on nutrient transfers from bush grazing to plots of land surrounding the village. However, over the last 30 years the area around Dilaba has gradually filled up, through extension of cultivated area, such that little or no fallow land now remains, all land receiving some fertilization when available. No distinction therefore exists between in-fields and outfields. For Dilaba, average yields were

950kg/ha for millet and 380kg/ha for sorghum. Yields of millet, which constitutes the main crop, are considerably higher for resource group I farmers than for those less well-endowed, the result of higher organic matter inputs from cattle holdings. Overall nutrient outflows were similar for all resource groups of household and field, and constitute a steady loss of N, P and K from the system. Resource group II and III farmers are trying to address these shortages by placing greater emphasis on fonio, cowpea, Bambara earthnut and sorghum, but must also rely on a range of other income sources such as migration and off-farm activities to make ends meet.

An important pattern across all sites, but particularly marked for Tissana and M'Péresso, was the major outflow of potassium. As was seen from the characteristics of soils for the four villages presented in Table 3.3, potassium levels are low in all cases and of potential concern especially in M'Péresso. Potassium outflows are generated by a large share of crop residue leaving the field system, through burning of stubble to make potash, collection as fodder, and grazing, without sufficient subsequent return of potassium through applications of manure or compost. These continued outflows are starting to receive attention from research and extension, with work on amending the mineral content of fertilizer used for rice and by investigating new ways of encouraging farmers to collect, compost, and recycle organic matter within their farming system. There are, however, several constraints on the latter option. In the irrigated rice sector, farmers recognize the potential benefits of composting residues, but find it easier to let cattle graze the stubble and burn off any remainder before ploughing. Equally, farmers in Dilaba were keen to use stubble for animals to graze, rather than collecting it. Only in M'Péresso was a more systematic harvest and composting of material carried out. This reluctance to invest a lot of effort in recycling of residues is in part due to limited labour availability. But it is also because crop residues are still thought of as a shared resource, so that once the harvest is safely stored, herds are usually left free to wander.

Overall, then, a number of patterns emerge from the nutrient-balance studies, in particular the marked contrasts between the cash crop and rainfed cereal zones. As expected, farmers in the cash-crop-producing areas of the Office du Niger and CMDT are much better able to maintain positive nutrient balances. This is due to much easier access to credit and agricultural inputs, combined with market prices which provide a sufficient incentive and means to intensify agriculture. By contrast, the cultivation of rainfed cereals – sorghum and millet – in Dilaba and Siguiné demonstrated persistently negative balances for all nutrients, a pattern which also held for cereal cultivation in cotton-growing M'Péresso.

ASSESSING THE SIGNIFICANCE OF THE PARTIAL NUTRIENT BALANCES

So what do these results tell us about broader trends of productivity and sustainability in these farming systems? At one level, there is cause for concern. The partial budgets, particularly for the cereal zones, provide support for the concern expressed in the policy statements presented at the start of this chapter. Particularly in the rainfed dry cereal villages, productivity is low and the level of nutrient extraction is high, indicating serious questions about longer-term sustainability, given low nutrient reserves in these soils. For the cash-crop zones, the balances are more favourable (except for potassium losses), with yields higher and nutrient balances being broadly maintained. The sustainability of these systems is highly dependent on attractive market prices, the continued availability of mineral fertilizer, an effective input supply system and access to credit.

Despite these broad generalizations, there are more fundamental problems in interpreting the nutrient balance figures. First, we need to remember that these are only partial nutrient budgets, and there are possibilities for other inputs and outputs beyond those measured. In the dryland cereal zone, this is particularly important in relation to dust deposition where significant amounts of nutrients can be added annually. Second, these are snapshot assessments, based on accounting for inputs and outputs over only one or two seasons. They do not take account of carry-over effects, which are very significant for organic applications, especially in the dryland zone, and fallowing sequences, which continues to be important in many villages like Siguiné.

In addition, it is necessary to look at the management practices which lie behind the figures, as these indicate both the potentials and constraints for positive change and improvement. Although the partial nutrient balances were often markedly similar across farmer resource groups, the similarity frequently hides considerable variation in terms of how people manage their soils. Thus, for example, all in-fields in Siguiné received a broadly similar input of organic fertilizer per hectare but, while for resource group I farmers this stemmed from cattle parked on the land, resource group III farmers had to rely on collecting and transporting household waste, a much more labour intensive task. In like manner, all farmers in M'Péresso ended up with broadly similar nutrient balances for cotton and for the cotton-millet rotation. But in this case, it was due to much larger volumes of organic fertilizer being generated by those in resource group I, while resource group III farmers depended more on mineral fertilizer and compost.

INSTITUTIONAL AND POLICY CONTEXTS: EXPLAINING DIFFERENCES IN SOIL-FERTILITY MANAGEMENT

In order to identify the potentials for improving soil-fertility management, and with this the productivity and sustainability of the farming system, we must look at the range of contexts which influence different farmers' ability to invest in their soils. As shown in the case studies, there are diverse and dynamic practices employed by farmers across the case study sites. The success or failure of these is affected by a range of institutional and policy factors which, in turn, affect access to factors such as labour, land, technology, information and skills, and markets and credit. It is to these institutional and policy contexts which we now turn, since strategies to address soil-fertility management need to recognize that farmers have markedly different patterns of access to such resources.

Access to labour: household structure and labour organization

The importance of access to labour is clear from the criteria used by farmers to classify good, medium, and poor soil-fertility managers. In the absence of a well-developed labour market, most households depend on their own members for work on the farm. Several of the case study farmers highlighted in Box 3.1 have found it difficult intensifying their farming due to shortage of family labour. Farmer A provided an exceptional case, in hiring 5 extra workers to cultivate his land. Farming households are often large and complex groupings of male kin with their wives and children. Management of the diverse interests, needs and assets of these groupings requires tact and good judgement, so that individuals are happy to continue to supply their labour, while benefiting from certain economies of scale and risk pooling (Brock and Coulibaly, 1999; Toulmin, 1992). Household heads need to balance the demands they can exert on young men's labour during the dry season for fetching and transporting organic materials to manure the fields, with the youthful desire to set off and earn cash on migration. In the case of Tissana such a balance can be neatly achieved, since individuals within the household have found that cultivation of irrigated vegetables provides a good source of dry season income. But elsewhere, a trade-off exists between migration for the individual's benefit and dry-season soil-management activities which contribute to the common household's welfare.

Malian farming households are often large and complex, with several generations living and working together as a single farming unit. Households with more than 50 members can still be found in many villages (Brock and Coulibaly, 1999). Many resources, including animal traction and cattle, are held in common by the household, although individuals may also own animals as separate property. The sheer complexity of labour mobilization and capital accumulation within these large groups makes household management an important issue. Traditionally, the head of the household, generally the oldest man, takes all decisions affecting agricultural production, investment and marriage plans. He may do this alone or in consultation with other male kin, and often delegates work in the fields to younger men. The productivity of the collective fields is influenced by the style of management and the capacity of the household head to motivate its members to work hard and invest in joint activities (see Box 3.2).

Large households face significant advantages, and economies of scale, such as an ability to mobilize capital from a range of activities, allocating labour to diverse income-generating sources in the long dry season, and providing greater protection from sickness and demographic risk than nuclear family groups (Toulmin, 1992). However, a conflict of interest or clash of personalities can initiate a process of household disintegration, with the cultivation of individual fields often the first step on the path of household

Box 3.2 Managing labour

B from M'Péresso has very little labour and, until recently, he had to manage without the help of his sons, who were both away on migration. However on their return, they brought back with them enough to purchase two oxen and a cart, thereby greatly improving the farm's capacity to cultivate land and generate dung. But both young men are likely to leave again for a long period because of the limited harvests produced on the farm. This richer neighbour sets an example to his workforce by leaving early each morning to go the field. He is always busy collecting materials to add to the compost heap and will pull up weeds, grasses and leaves from communal areas to add to the heap. But he also gives his children money to help him with the farm.

Successful household heads need to generate a strong spirit of cooperation and joint enterprise amongst family members, particularly when they have few assets. K is amongst the poorest in Dilaba, yet he and his family are considered by all who know them a really hard-working team. He, his wife and their single daughter get up early and work all day through – other villagers say 'they're always in the field' – producing millet, fonio, groundnuts and bambara earthnuts. Both he and his wife invest extra time in fattening sheep for market, using what crop residues they have available as well as bran from pounding cereals into flour. They also were the first to experiment with the chaff cutter brought in by researchers to improving crop residue palatability. Their evident industry makes it easier for them to ask for help from others. In 1996, one of his oxen died, but he was lucky enough to be able to borrow one from his wife's parents. He owns a donkey cart which he lends out to a neighbour in exchange for help at harvest time. He also has a friendly relationship with a richer farmer in the village to whom he can turn for help in times of need.

division and separation. However, in the rice irrigation scheme some families break up on purpose, as this may improve their possibility to gain access to new rice fields through the Office du Niger.

The household usually farms a series of collective fields to which both men and women are expected to devote themselves. Individuals may also have their own fields, such as the small plots women farm after they have finished work or on their free days. Older women who have retired from household chores have greater time available to work their fields and can produce a more substantial harvest. Equally, retired men usually farm a small plot of maize, tobacco or groundnuts. In the irrigation scheme, household heads may allow individuals to cultivate vegetables to earn their own cash in the dry season.

The division of labour in rainfed agriculture has changed over time, given the influence of increasing use of animal traction for land preparation and weeding, which tend to be male tasks, while women come to the fields once they have completed domestic chores. In the cotton zone, the CMDT has been putting more emphasis on the quality of cotton fibre, which has led to more labour-intensive harvesting methods (Dembélé et al, 1997). The division of tasks by gender is more distinct for irrigated rice cultivation, where men are responsible for land preparation, weeding and harvesting. Women are mainly involved in transplanting rice seedlings and the collection of harvested rice. Children are also heavily involved in agriculture and in herding of livestock. For this reason, many households, particularly in more remote areas, have not been keen on sending their children to school since they consider this will deprive them of an important source of labour during school terms, while offering no prospect of improved job opportunities.

Access to land: land tenure and management institutions

Outside the Office du Niger, access to land is largely organized through customary land tenure systems with farmers holding usufruct rights. The family that founded the village usually takes the role of *chef de terre* and is responsible for resolving conflicts over land. Most families do not perceive land tenure as insecure, although recent immigrants may feel less at ease. However, in many areas, access to land for future generations is increasingly perceived as problematic, as there is less and less good quality farmland available. Access to land within the irrigation scheme is organized by the Office du Niger through leases to tenant farmers. Recent changes in terms of the lease now provide much greater security with plots being transmissible to heirs (Jamin, 1995).

While rights to farm land are considered relatively secure, rights to control access to common property resources within the village's territory are much less clear. Under current legislation, the Malian government is the owner of all land that has lain fallow for more than 10 years, as well as woodland and grazing areas. As a result, villagers are not able to exert effective control over who uses the pastures and woodlands around their settlement. Increasingly, local people want to assert such rights to manage these resources, which are vital to the sustainability of farming and livestock systems, as well as being a valuable source of bush fruit and fuelwood (PADLOS, 1998; Dème, 1998). The village of Dilaba faces particular pressures from outsiders seeking hay from their common grazing areas, which are already in limited supply. A number of pilot agreements have been developed which authorize local villages to take up firmer management tasks and impose controls and fines on the unauthorized entry and use of their lands. For example, M'Péresso is part of such an association, called the Siwaa, which has been trying to get government recognition of its right to exclude outsiders and establish its own by-laws regarding access to and use of common resources (Joldersma et al, 1996; Hilhorst and Coulibaly, 1999). New legislation being discussed aims to clarify the division of responsibility for such resources between village associations and the newly-created communes elected in May 1999 under the decentralization process. However, it is as yet unclear how the potential overlap of responsibilities will be resolved in practice.

Access to technology, knowledge and skills: research and extension approaches

The experience of the cotton sector shows the valuable role which can be played by research and extension in improving soil-fertility management.

Farmers in M'Péresso have been linked to the Sikasso research station for the last decade, during which time new ideas have been introduced and spread amongst many farmers. This willingness to test out improved measures – such as recycling of cotton stalks through cattle pens and setting up composting systems – shows the readiness of farmers to adopt potentially beneficial techniques. Thus, for example, all farmers in M'Péresso generate very substantial quantities of organic materials to fertilize their fields, and far in excess of those produced in the other three villages. The rice system of the Office du Niger has also benefited from considerable investment in research. Once incentives and production conditions were improved, with the rehabilitation of the scheme in the early 1990s and the devaluation of the CFA franc, farmers rapidly took up new technologies.

However, research and extension support has tended to concentrate on one crop or element of the farming system. Yet, for example, cash-crop farmers are also engaged in many other activities. Rainfed cereals play a major role in the cotton system, while rice farmers are increasingly investing in irrigated vegetables, as well as having a herd of cattle and a dryland millet field. At the same time, the farmers of Siguiné are starting to gain access to irrigated land on the edge of the Office du Niger, which is in the process of expansion. Thus, in future, Siguiné's farmers may become much more dependent on irrigated agriculture, and leave much of their dryland farming behind. Farmers in Dilaba are the least able to generate a range of options by which to diversify their farming activities because of the limitations which they face, as discussed in more detail below. This diversity in the patterns by which people gain incomes means that research to support farmers in villages like Dilaba needs to move away from a focus on recommendations linked to a single crop, towards understanding the linkages between different components from which people construct their livelihoods (see Box 3.3).

As the case studies have amply shown, farming populations in Mali are far from uniform. Different groups – differentiated by wealth, age, gender, ethnicity etc – have different asset bases, interests and capacities, and incentives to invest in soil management. Too often research and extension focuses on the 'average' farmer, and ignores other groups. Poorer farm households in the rainfed villages of Dilaba and Siguiné face particular difficulties, and could usefully benefit from creative thinking from research and extension services. Extension recommendations which focus on increasing the production and

BOX 3.3 DIVERSE LIVELIHOOD STRATEGIES

J in Siguiné is better-off than many, and leads a large extended family. While he concentrates on farming, his two brothers work part-time as a builder and blacksmith respectively, and they invest such revenues in cattle holdings. They are well equipped with more than 20 head of cattle, as many sheep and goats, 2 ploughs and 2 carts. The cattle herd is particularly important as a source of cash in years of poor rainfall, which are frequent in Siguiné. They also provide a means to pay the costs of marriage.

recycling of organic and waste materials have a heavy cost in terms of the labour input required. Resource group III farmers tend to be poor, not only in terms of livestock and other assets, but also access to and control over labour (as was seen from Table 3.3). As a result, they do not find it easy to mobilize the labour and transport needed to carry crop residues, household waste, and compost materials between homestead and field. Thought needs to be given to ways of helping such farmers by minimizing transport requirements. Equally, selective use of small quantities of mineral fertilizer and careful placement may be of significant value. In 1997, several farmers in Siguiné had started buying small quantities of DAP fertilizer to mix with seed. Such a practice could usefully be spread and improved through experiments with different amounts of fertilizer, or varying blends, to assess ways of maximizing its value, particularly in combination with organic materials.

The value of participatory research approaches, combining the knowledge and insights of farmers, extension agents and researchers, was made evident during the course of this research. In all four villages, debate between farmers and researchers was greatly aided by the use of resource-flow maps, drawn jointly by farmers and research staff and providing a focus for discussion of current practice and potential for improvement (see Chapter 5). This led to the testing out of several new measures in each case, aimed at addressing the constraints faced by farmers in different resource groups. These included use of compost pits next to the field to minimize problems of transporting organic materials, use of bedding materials to improve the volume and effectiveness of animal manure, and the introduction of a cutter to prepare millet stover for animal fodder. The regular presence of researchers in itself seems to have spurred increased reflection on ways to improve soil management.

BOX 3.4 FARMER EXPERIMENTATION IN SIGUINÉ

A member of resource group III, N in Siguiné decided in 1997 to experiment with small amounts of di-ammonium phosphate (DAP), the most commonly found inorganic fertilizer on the market. N is someone who travels a lot, and spends a considerable time in Niono, the neighbouring market town and administrative centre for the Office du Niger. He mixed the DAP with seed at sowing time and was very pleased by the results. In 1998, several other farmers followed his example.

Support from research and extension staff can have an effect on farm management practices. When visits began to Siguiné in 1996, there was an enormous heap of household waste on the edge of the village where everyone dumped their rubbish. During feedback sessions between villagers and the research team, the potential value of the waste heap was discussed, and by the next year the heap had entirely disappeared to be spread onto the fields. Each household has now established their own compost heap for household waste.

Access to inputs and credit: institutions for marketing and credit supply

Access to inputs and credit is very different between the cash-crop zones and the dryland cereal areas. In the cash-crop zones, especially in the CMDT area, a structured system of credit provision and input supply is combined with a well developed marketing system supported by good infrastructure. For many years, the CMDT has provided cotton farmers in Mali with a stable structure of guaranteed prices, input quality and timely delivery which has encouraged continued investment by farmers in cotton production. At the same time, they have maintained responsibility for support to the broader farming system, whether food crops, livestock, or environmental management. The CMDT has, however, come in for serious criticism regarding the profit margin they reap by keeping the price paid to farmers low, while charging a high price for the various inputs and services they provide the farmer. The cotton farmers' union, SYCOV, has played an important role in challenging and negotiating price levels, and ensuring farmers get a somewhat better share of the world market price. This escalated into a major cotton farmers' strike during 1999–2000. As noted earlier, the CMDT appears to have withstood the pressures upon it to be privatized in return for promises to develop a more transparent system of management and accounting.

Micro-finance institutions in Mali are mainly concentrated in the southern region of the country. Credit networks, such as *kafojigine*, provide one means to mobilize savings and provide credit for a variety of purposes, but are not widespread as yet outside the CMDT area. However, in CMDT areas, indebtedness has become an increasing problem. For example, farmers may take a loan for purchase of mineral fertilizer but then use it for some more pressing purpose, and have only a poor cotton harvest with which to repay the loan. Equally, since 1995, an increasing number of credit suppliers have been setting up in southern Mali, providing access for AVs to loans for a wide variety of purposes using the cotton harvest as collateral. In some cases, such loans cannot be repaid and several credit suppliers may find the cotton harvest pledged two or three times. The CMDT is then usually obliged to step in to bail out the AVs which have got into debt.

By contrast, the Office du Niger has lost many of its former powers over rice farmers and no longer has monopoly rights to buy the rice harvest. Equally, provision of credit, extension advice and input supply have all been removed from their mandate. As a result, farmers have less assured access to these services. In the Office du Niger, a number of AVs have become non-credit-worthy as a result of defaults, but have recently been given a reprieve following intervention by donors. For farmers who cannot gain credit from their AV due to outstanding debts, reliance on traders has become common not only for the inputs needed for rice production, but also for vegetables, with the development of contract farming, especially for tomatoes. AVs in the Office du Niger are also trying out new ways of coping with farmer debt. In some cases, the AV takes over part of the farmer's irrigated holding and will manage these until the debt is repaid. The farmer is left with enough land to provide for his own food.

No such formal credit and extension structures exist for dryland villages like Siguiné and Dilaba, although occasional opportunities present themselves through project-based initiatives by which means dryland cereal farmers can gain access to credit, as was the case for farmers in Dilaba, with the establishment of the IFAD-funded PFDVS. As a result, farmers must rely on capital generated by the household's own resources rather than taking out a loan. Migration earnings are a particularly significant source of capital for farm investment, while social networks also provide informal sources of credit.

CONCLUSION

If farmers are to seize the opportunities for improving soil-fertility management, the broader conditions must be right. Thus institutions and policies governing access to labour, land, technology and knowledge and markets and credit are central to any strategy for the improvement of soil-fertility in Mali. While there are clearly important technical issues to be addressed (such as the large outflows of potassium in a number of sites), these must be understood by looking at the broader context of how farmers' livelihoods and farming practices have been changing in the diverse settings found in Mali.

As the case studies have shown, farming systems in all sites are highly dynamic. Over time, farmers in each of the study villages have been adapting substantially the way in which they farm and their associated soil-fertility management practices. For example, in villages like Tissana in the irrigated rice sector, yields have risen greatly as a result of better water control following the rehabilitation of the irrigation system's infrastructure, rising prices and the use of new technologies. Equally, many farmers are now starting to produce vegetables in the dry season, for which they must generate larger amounts of organic fertilizer. Such diversification is likely to continue. Siguiné's farmers are now investigating the opportunities for irrigated rice cultivation to supplement their high-risk rainfed millet farming system. In Dilaba, the situation faced by poorer households is more problematic. The occupation of all available farm land has led to a growing scarcity of grazing and possibilities for maintaining nutrient flows from common pastures to cropland. Farmers face a pressing rising need to find alternative sources of income, such as a more diverse spread of crops, including cowpea and Bambara earthnuts. The experience of cotton farmers in M'Péresso, and the CMDT zone more generally demonstrates how the combination of good economic returns and available technologies can generate a strong incentive for farmers to invest in maintaining soil-fertility. Cotton production has boomed in the years since the 1994 devaluation, because world market prices were, initially, also rising. However, with prices now faltering at a global level, the current pattern of investment in large inputs of inorganic and organic fertilizer is unlikely to persist.

Changes in farming practice must also be seen in a wider livelihood context. Farming is one among a number of activities in which rural people are engaged, and understanding the role of soil-fertility management in this broader livelihood portfolio is key if interventions, whether focused on technical issues at the farm level or broader institutional or policy levels, are to have an effect. Diversification within crop and livestock activities is a common feature in high-risk areas, and is but part of the broader diversification of livelihoods and incomes found amongst Sahelian populations as a whole. Thus, many farm households aim to combine a varying array of earnings from cultivation, livestock herding, migration, craft, petty trade and transport of goods. For example, diversification of livelihoods out of agriculture is very important for Dilaba families, with migration playing a major role in making ends meet. In M'Péresso there is a continuing integration between crops and livestock. While cotton provides reasonable returns for many farmers, as seen by the gradual extension of the area devoted to this crop, a number of farmers are also moving into other crops such as maize and vegetables.

Farmers in Mali, across very different settings, have an active interest in trying to improve the fertility of their soils, whether they be in cash-crop areas or in marginal cereal zones. The effort they devote to such activities depends not just on access to technologies, but also on returns to agriculture in relation to the many other opportunities they face. Despite harsh and difficult conditions, both climatic and economic, Mali's farmers have shown much ingenuity in seizing new opportunities for diversifying their livelihoods on and off the farm. Research, extension and policy design need to recognise such dynamism and diversity in practices followed and constraints faced, to ensure that national strategies provide an enabling framework for learning from the field and support to local-level experimentation.

Chapter 4

SOILS, LIVELIHOODS AND AGRICULTURAL CHANGE: THE MANAGEMENT OF SOIL-FERTILITY IN THE COMMUNAL LANDS OF ZIMBABWE*

C Chibudu, G Chiota, E Kandiros, B Mavedzenge, B Mombeshora, M Mudhara, F Murimbarimba, A Nasasara and I Scoones[1]

INTRODUCTION

Policy debates about soil management in the communal areas of Zimbabwe date back to the early colonial era when soil erosion and soil-fertility loss were deemed to be undermining the productive base of the country (Beinart, 1984). Since that time much scientific effort has been invested in documenting the extent and nature of the problem through a string of field surveys, mapping exercises and biophysical and economic modelling efforts (eg Elwell, 1974, 1983, 1985; Whitlow, 1988, among many others). The result has been the emergence of a conventional wisdom which states that soil erosion and fertility loss is bad and getting worse. This view is so often repeated in both the scientific literature and in the popular media that its basic premises are barely questioned.

This particular view of communal area soils has informed much development policy and intervention over the past 60 years or more. Since the 1930s, and particularly following the enactment of the Natural Resources Act of 1942, the prevention of soil erosion has been at the centre of agricultural development efforts. Indeed, the major attempt at land reorganization and agricultural modernization promoted by the Native Land Husbandry Act of 1952 was particularly focused on soil management (Floyd, 1961). Contour ridging, drainage channels and tillage systems have been at the centre of agricultural extension efforts ever since (Vogel, 1994).

In recent times a renewed concern with environmental management and soils in particular, has been apparent. A number of rather alarmist statements about the prospects for calamity have prompted this increase in government and donor interest. For instance, the respected scientist Henry Elwell commented:

> *There is absolutely no doubt whatsoever that malnutrition and death through starvation of the communal land population is inevitable if present rates of soil erosion are allowed to continue.* (Elwell, 1983, p145)

This refrain was in turn picked up in wide-circulation policy briefings supported by the Southern Africa Research and Documentation Centre (SARDC, 1994). The agricultural extension department's (Agritex) manual on soil conservation and a leaflet aimed at the interested public produced by the Department of Natural Resources present a similar picture.

Our concern is not to dismiss such statements out of hand. Rather our interest is to interrogate them critically, offering perhaps a somewhat more nuanced perspective on soils management than has hitherto been given. Such a perspective requires looking in detail at what is going on in particular places with particular people, asking where soil-fertility is declining (or improving) and for whom? This, in turn, requires looking at how farmers themselves understand their soil resources and the dynamics of change, examining historical changes in agriculture, soils and landscapes. Our concern, therefore, is not simply with the technical details of soil loss or nutrient balance, but with the social and ecological contexts for changes in the soil resource over time. Not surprisingly our results show a rather more complex story than the simplistic stories promulgated as part of the technical and policy debate over the past 60 or so years.

Following a brief review of some of the major research directions on soils management, the chapter turns to an examination of data derived from a detailed examination of two case study sites located in Mangwende and Chivi communal areas. Through a comparison between sites and among farmers, insights into the diversity and dynamics of soil-fertility management practice are offered. An examination of nutrient balances across different agro-ecological sites and socio-economic groupings highlights a range of issues concerning soil-fertility dynamics and sustainability. This discussion, in turn, leads into an exploration of the social and economic contexts of soil-fertility change. The range of soil-fertility management practices are explored and biographical case studies are used to examine how particular social, economic and policy factors influence the way soils are managed by different people in different places. The final section of this chapter briefly examines some of the implications of these findings for policy and practice.

RESEARCHING SOILS IN ZIMBABWE: A BRIEF OVERVIEW

A considerable body of formal research on soil-fertility and its management exists for Zimbabwe, dating back to the early part of the colonial era. While much of the early work was directed towards the emerging white-settler commercial farming sector, nevertheless scientific understanding of soil properties and the responses of crops to different combinations of inputs is extensive. This major investment in scientific work can be characterized in a number of phases. These are briefly explored below.

The earliest experiments established on the new research station in Salisbury (now Harare) focused on methods for maintaining yields in continuously cropped maize, as newly arrived farmers from Europe found that yields quickly dropped in their fields cleared from the bush. A series of rotational experiments were established in 1913 and ran until the late 1950s. Such experiments, although with different treatments, were extended to other stations, including Marondera, Makoholi and Matopos, during the following decades. The experiments showed that if rotation of inputs and fallowing occurred, yield levels could be maintained significantly above the control plots where only maize was grown with no inputs (Saunder, 1960).

Over the years, research attention was focused on different inputs into the rotation. Composting and green manuring approaches became particularly popular in the 1930s, drawing inspiration from earlier work in India on composting and green manuring. Between 1933 and 1940, 20 articles in the *Rhodesian Agricultural Journal* were devoted to composting, while only 6 appeared in the subsequent 25 years. Similarly, publishing on the theme of green manuring was concentrated in the same period, with 10 articles published from 1929 to 1940.

A mixed farming model emerged as the ideal solution, whereby manure would be generated by livestock and would, in turn, be the basis for sustaining crop production, as part of a rotational system which would include the growing of cover crops for green manure and leys for fodder. The missionary, and later Chief Native Agriculturalist, E D Alvord was a firm advocate of this approach (Alvord, 1948), and, following experimentation in Shurugwi, encouraged others to adopt his recommendations (Wolmer and Scoones, 2000). These involved the application of 37 tonnes of manure per acre in the first year, followed by a four-year crop rotation. Subsequent research at the research stations at Grasslands (Grant, 1967 a, b, 1970, 1976) and Henderson (Rodel and Hopley, 1969, 1970; Rodel et al, 1980), as well as in the communal areas (Grant, 1981) on manuring (and combinations with inorganics) has adapted these recommendations to suggest that a 10-tonne/ha application each year is more appropriate (Grant, 1981). However, the basic mixed-farming model reliant on large quantities of manure has remained and has continued to inform much of the technical research since the 1930s.

The adoption of consolidated arable units as part of the mixed farming approach was linked to a parallel debate about soil conservation and the need for mechanical conservation measures combined with sound land-use planning to combat perceived threats of environmental degradation and desertification. Concerns about soil erosion reached their height in the 1930s in southern Africa, particularly following the Drought Commission in South Africa in 1922 (Beinart, 1984). Various reports in Zimbabwe confirmed the same fears that continued soil erosion would soon undermine the productive base of the colony (eg Aylen, 1950). This broader concern was responded to, and fuelled by, a scientific and technical discussion. Some 26 articles on soil erosion and conservation were published during the 1930s in the *Rhodesian Agriculture Journal*, more than on any other subject relating to soil management.

The policy concern with soil erosion led to the initiation of a number of trials, including the famous Henderson erosion plots where work was carried out by Hudson between 1953 and 1963 (Hudson, 1962). Other similar plots were also established at Makoholi and Matopos research stations. These studies confirmed the potential for very significant soil losses from arable plots. Later, building on this work, attempts at developing a predictive model for estimating soil-loss levels were made (Stocking and Elwell, 1984). Surveys of both potential and actual erosion were carried out in the 1970s and 1980s when concerns about environmental degradation were again heightened (eg Elwell, 1974, 1983; Elwell and Stocking, 1988; Whitlow, 1988), and estimates of losses of 50 tonnes of soil per hectare from arable land were widely claimed.

Through this period the Institute of Agricultural Engineering at Hatcliffe near Harare became an important centre for research on soil-erosion issues, conservation tillage, and agricultural technology. During the 1970s interest in conservation tillage expanded in the commercial farming sector but was also pursued in the communal areas, particularly following independence in the context of the Agritex/GTZ Contill project based at Makoholi research station near Masvingo. Extensive work was undertaken on different types of tillage techniques (Chuma and Hagmann, 1995; Hagmann, 1995; Vogel, 1992, 1994).

Investment in soil survey and soil- and land-capability classification became an important part of efforts by the colonial government to plan land use, particularly in the 'African reserves'. This reached its height in the 1950s with the implementation of the Native Land Husbandry Act, although planning for centralization in the reserves preceded this, and similar approaches continue to be used by rural planners and extensionists today. To complement such efforts, soil surveyors designed a locally appropriate soil-classification system (Thompson and Purves, 1981) which was widely used. Broader agro-ecological classifications into regions of different potential (Vincent and Thomas, 1962) also made extensive use of soils research.

From the 1950s technical research on soils management changed dramatically, with the increasing importance of mineral fertilizers associated with the hybrid maize revolution being promoted across the country, including in the communal areas. Experiments on mineral fertilizers were established on all research stations from the 1950s and these were regularly reported in the local journals, with endless elaborations of different response curves under a wide range of conditions and test crops (eg Metelerkamp, 1988). More recently attention has extended beyond the agronomic details to a consideration of the economics of fertilizer use, especially in the communal area setting (eg Mataruka et al, 1990; Shumba et al, 1990; Mudhara, 1991; Page and Chonyera, 1994).

Following independence, official research priorities shifted towards supporting communal area agriculture. Soils research reflected this change, with a range of surveys highlighting the issues in communal areas (eg Shumba, 1985; Campbell et al, 1998; Carter et al, 1993; FSRU, 1993; Campbell et al, 1997). The general assumption that soil degradation is proceeding apace in the communal areas was reinforced by a number of influential surveys looking at erosion and land degradation (Whitlow, 1988; Whitlow and Campbell, 1989).

An increasing recognition that inorganic fertilizers were unlikely to solve the problems of soil-fertility in the communal area sector, particularly among poorer farmers and in the drier areas, has redirected attention back to organic sources. Post-independence research, while maintaining a continued interest in inorganic fertilizer applications, has again returned to looking at issues of manure use and quality (Mugwira, 1984; Mukurumbira, 1985; Mugwira and Mukurumbira, 1984, 1986; Mugwira and Shumba, 1986; Khombe et al, 1992), green manures and composting (Hikwa et al, 1997), nitrogen-fixation potentials (Shumba and Dhliwayo, 1990; Mapfumo et al, 1998), and the potentials of organic/inorganic mixes (Trounce et al, 1985). There has also been some important work on soil biological and chemical processes which has illuminated issues surrounding the processes of decomposition and mineralization of different materials, for example (Murwira et al, 1995; Murwira, 1994, 1995; Murwira and Kirchmann, 1993a). With the recognition that efficient use of limited materials is key, there has been some work on efficiency, synchrony and release dynamics (Piha, 1993; Murwira and Kirchmann, 1993b), as well as placement (Munguri et al, 1996).

Thus, over time, different emphases in the technical research are evident, with shifts from an organic-matter-management focus in the 1930s and 1940s to a concentration on inorganic fertilizers from the 1960s. Today, a more integrated stance is evident, with a growing recognition that it is the efficient application of a combination of fertility resources that offers the highest potential returns (Carter and Van Oosterhuit, 1993; Kumwenda et al, 1995; Hikwa and Mukurumbira 1995; Hikwa et al, 1997; Giller et al, 1998). But much of this research has focused only on the plot level, exploring how particular crops respond to particular input combinations. While such insights are important, they often fail to situate results within a broader understanding of farming system change. They assume – often only implicitly – that the desired end-point is a particular type of 'mixed farm' model, where soil-fertility decline is offset and crop productivity assured by the integrated management of crops and livestock and the application of manure and fertilizer. The following sections will examine the diversity of farming system types, differentiated both by site and farmer, and the degree to which one standard pathway of change is evident. First, however, some characteristics of the case study sites need to be introduced.

THE CASE STUDY AREAS

The case study areas represent two contrasting communal area settings (Figure 4.1). Mangwende, some 60kms from Harare, is in a relatively high-potential zone, where rainfall averages 820mm per annum (CV [coefficient of variation] 25 per cent) in Mrewa town. Maize production dominates the agricultural system, although significant incomes are also made from vegetables which are marketed in Harare. Non-arable areas are characterized by a *miombo*-type vegetation with *Brachystegia spiciformis* and *Julbernadia globiflora* being the dominant trees. Soils are largely poor, granitic sands with limited available nutrients, although valley bottom *dambos* tend have richer alluvial soils.

Figure 4.1 *Map of Zimbabwe and location of study areas*

By contrast, Chivi is in the drier southern part of the country, where rainfall averages 550mm per annum (CV 37 per cent) in Chivi Office. Again, maize is the dominant arable crop, although yields are considerably lower and droughts often mean that no surplus is produced. Maize is combined with a range of small grains, including pearl and finger millet and sorghum, as well as some key cash crops such as cotton and sunflower. The major vegetation types include patches of *miombo* woodland, but also *mopane* woodland more typical of the drier low-veld. Soils are generally more fertile than in Mangwende, although there are extensive granite sand areas where long-term cultivation has reduced fertility. Such areas, though, are interspersed with heavier clays, both in the valley bottoms where *dambos* may be important, but also – more extensively – the result of doleritic intrusions.

In both areas, livestock are an important part of the farming system, and critical to understanding soil-fertility dynamics, due to the importance of manure. In neither area are livelihoods completely reliant on agricultural production, although this is more feasible in the higher-potential area of Mangwende. Remittances from urban employment are critical, both for day-to-day survival in the rural areas, but also for investment. Some of the major contrasts between the two sites are highlighted in Table 4.1.

In order to investigate soil-fertility issues in detail, two sites were chosen to be broadly representative of the communal area as a whole. A number of research activities were carried out in both sites between 1994 and 1998, including participatory rural appraisals, soil surveys and sampling, baseline household questionnaires, key informant interviews, historical interviews, nutrient-budget analyses and participatory farmer-led experimentation on soils. The following sections report the results of this research, highlighting

Table 4.1 *Some contrasts between Mangwende and Chivi communal areas*

	Mangwende	*Chivi*
Rainfall (mm/year average)	820 (CV 25 per cent)	548 (CV 37 per cent)
Altitude (masl)	1200–1500	600–1300
Major soil types	Granite sands, mostly of the paraferrallitic group. Some small areas of doleritic clays; low to medium fertility sands, many deficient in N; low water holding capacity; low pH	Granite with banded ironstones; most kaolinitic and ferasalitic sandy loams; some patches of doleritic clays; deficient in N and P; low water holding capacity; low pH
Population density (people/km^2)	97	44
Farm size (ha: range from poorer to richer group average)	1.8–4.1	2.8–4.4
Cattle holding (N: range from poorer to richer group average)	0–10.0	1.4–7.5
Inorganic fertilizer use (% of households applying to outfield)	66	16
Major crops	Maize, groundnuts, vegetables, some cotton	Maize, sorghum, pearl millet, groundnuts, sunflower, some cotton
Distance from major towns	Mrewa within communal area; Harare 60km from Mrewa	Masvingo c.50km from Chivi Office and 100km from Ngundu; Harare 350km; Bulawayo 150 km

key aspects of the methodologies used where appropriate. It is worth pointing out at this juncture the important synergies that developed between the different elements of field research and action which contributed to our learning and insight. None of the single elements alone would have provided the type of understanding of the different settings that has emerged. For example, the nutrient-budget analyses required information from environmental histories, social biographies over several generations and the questionnaire surveys in order to interpret the results. Similarly, questions raised by the farmer-led experimentation and action research were key in pushing us to investigate aspects of soil management further, uncovering local technical innovations and management practices which would have remained hidden.

But before examining these issues in detail, some further background on the case study sites is needed to set the results in context. For, as will be seen below, it is the historical trajectories of change in agricultural practice and social, economic and institutional settings which influence current conditions.

ECOLOGICAL DYNAMICS AND SOIL-FERTILITY

The two research sites represent a range of ecological conditions, with important implications for the dynamics of soil-fertility management (see Chapter

1, Table 1.2). Mangwende is a typical dystrophic savanna setting, with relatively high rainfall and poor granite soils making nutrients (and particularly nitrogen) a limiting factor to plant and crop growth in most years. Indeed in some parts of the landscape, notably the low-lying valley bottom, water may be in excess and strategies for reducing waterlogging must be employed. However, in such sites nutrients are unlikely to be limiting, as these areas act as nutrient sinks capturing eroded soil, leached nutrients and organic matter washed from the toplands. By contrast, parts of the Chivi area can be characterized as a more eutrophic savanna type, where water rather than nutrients are limiting to primary production. This is particularly the case in the heavy soils which are scattered through the area. In the other sandy areas more dystrophic conditions apply, although water may still be limiting, especially in the regularly occurring dry years. Water limitation is only overcome in low-lying areas, alongside rivers or in valley bottom lands. Here high levels of plant-available moisture can often be found year round, and crop growth may even be possible in the worst of droughts.

A number of studies have been carried out to investigate which nutrients are limiting to crop growth in the communal areas. Grant (1981) reviews some of this data, focusing on the typical sandy soil types. Overall nitrogen is the key limiting nutrient, especially in the heavily-leached sandy soils where organic matter content is low. The low organic matter content of soils also has an impact on sulphur levels, and this may be a problem in some areas. Phosphorous may also be lacking in such soils, especially as it has been found that manures appear to offer insufficient P for effective plant growth, even if applied in large quantities. Potassium by contrast has been found to be rarely limiting, and is generally not considered in work on soil nutrients in Zimbabwe. However, a range of trace elements – including zinc and magnesium – have been found to be important in particular places, although most research fails to investigate these in detail.

Within the broader classification of savanna type which differs between the two sites, other smaller-scale spatial variations occur within the landscape, where different limiting conditions of available nutrients or moisture may apply. Such patchiness has important implications for soil management and agricultural potential. The distinction between the drier toplands and the valley bottoms has already been noted. Such spatial patterning imposed by natural variations in topography is complemented by other sources of spatial variability in the soil landscape. These may be the result of past human action, such

Table 4.2 *Soil test results from clay and sandy soils in Mangwende and Chivi*

	Clay (%)	*Cation Exchange Capacity (meq %)*	*Organic C (%)*	*pH*
Mangwende – clay	31	28.0	0.78	4.7
Mangwende – sand	10	1.1	0.54	4.2
Chivi – clay	43	31.0	1.37	6.4
Chivi – sand	10	2.9	0.24	5.5

as the nutrient hotspots created by past settlement or *kraal* sites, or part of on-going differential transformation of soils through agricultural activity, as with the differences in soil quality between homefields and outfields (see below). Such human influences may interact with biological processes, such as the establishment of termite colonies and the creation of mounds. Different termite species are attracted to different soil types and levels of clearance and disturbance. Thus, depending on how the environment has been changed through human action, a different termite fauna is likely to be present. The presence of termites, in turn, influences the spatial patterning of soil quality, with termite mounds being highly valuable concentrations of high-quality clays and other minerals in an otherwise sandy landscape (Watson, 1977). The presence of certain types of trees similarly has a localized effect on soil quality. Leaf litter fall combined with the deposition of animal dung around large trees, where animals are attracted to shade, results in an enhancement of soil-fertility levels around such trees (Chivaura-Mususa and Campbell, 1998; Campbell et al, 1988). In both Chivi and Mangwende, large fruit trees (eg *Sclerocarya birrea*, *Ficus* spp, *Parinaria curatellifolia*) are kept in fields for fruit, shade and soil ameloriation (Wilson, 1989).

Thus across these areas a spatially-variegated mosaic of different conditions applies, with spatial variation occurring across slopes and as part of a patchwork of distinct niches within a broader soil landscape. This mosaic in turn varies over time, both between seasons and years. Ecosystem dynamics have changed significantly over the last 100 years or so, as land has been cleared for agriculture, livestock populations expanded and settlements established. In the past, in the *miombo* ecosystem (Frost, 1996), a combination of elephant damage and fire were important drivers. Within the larger landscape a relatively stable system existed, although at smaller scales complex patch dynamics were important factors in nutrient cycling at particular sites. But, as wildlife populations declined and human populations expanded, a different dynamic has emerged. With permanently cleared fields the number of trees in the landscape has decreased and, with this, shifts to other vegetation states have occurred, including other woodland types and grassland. This encouraged more regular dry-season fires, resulting in an increasingly pulsed pattern of nutrient release. This was reinforced by the persistence in some areas of a form of shifting cultivation involving regular burning. Over time, though, fire frequency has declined as more areas became permanently occupied and cultivated, and the growing livestock populations consumed a larger portion of the available biomass. In cleared areas a number of changes can be observed, including a change in the spatial patterning and temporal dynamics of the system resulting from shifts in the composition of both vegetation and soil fauna and, with this, changes in the nutrient cycle (Campbell et al, 1997). Comparisons of *miombo* and arable sites shows that permanent cultivation results in a decline in the availability of organic carbon in the surface layer (King and Campbell, 1994). This is due to the decrease in litter fall and the export of crop residues from the field. With limited external inputs, a much more patchy distribution of nutrient availability is the result, with areas below single trees or termite mounds

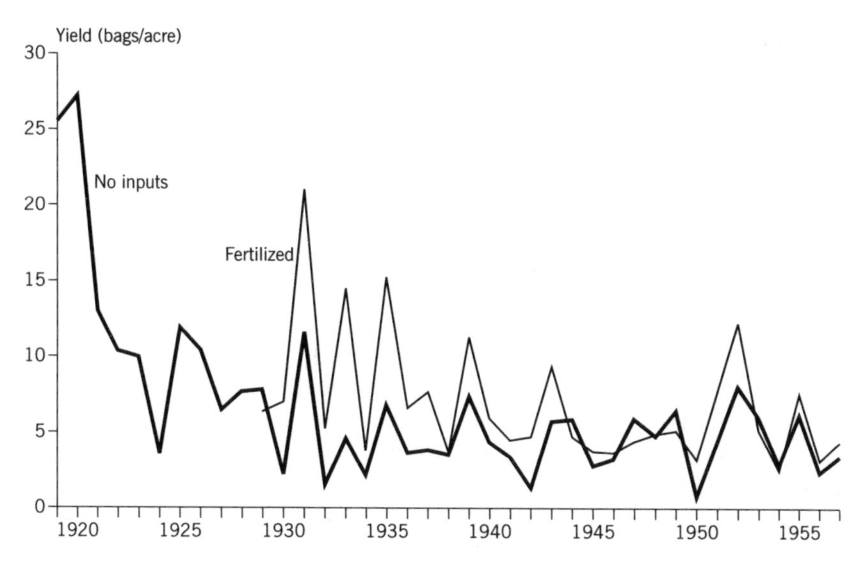

Figure 4.2 *Time series data of maize yields 1919–1957, with and without inputs (Harare Research Station)*

being the site of significant nutrient concentrations in what is otherwise a generally very poor sandy soil.

Long-term experiments which monitor the changes in soil properties and crop yields following clearance in environments of this sort highlight a number of key features. Figure 4.2 shows one example of a time series of this sort derived from the Harare Research Station. First, yields decline quite quickly following clearance, especially if no additional inputs are added. Second, whatever inputs are added, the yield levels realized in the first years after clearance are rarely attained; highest yields tends to come from mixes of organic and inorganic inputs. Third, soil-fertility decline may not be due to declines in macronutrients alone, as in some experiments neither total nor available N declined over time (eg Grant, 1967a). Under conditions of continuous cultivation it seems three different outcomes are therefore possible (see Chapter 1). Either yields decline to a low-level equilibrium which varies depending on rainfall, but very little output is achieved; or yields are raised from the low-level equilibrium and maintained through a regular input of nutrients to alleviate key constraints; or a longer-term investment in soil-fertility is made which boosts productivity to a higher state which can be sustained with limited maintenance inputs. Which of these three options emerges in any particular place, however, will be affected by temporal dynamics, among other factors.

Temporal variability is particularly apparent in the drier site of Chivi, where variations in rainfall are especially significant in ecological dynamics. A particularly important seasonal characteristic of these areas, particularly in the *miombo* woodland areas with sandy soils, is the nutrient flush which occurs prior to the first rains (Frost, 1996). This is the result of rapid mineralization on the wetting of previously dry soil. It means that dry planted crops can make use of the relatively high availability of nutrients at the same time as the first rains, capturing the limited nutrient sources available in these poor soils with high levels of efficiency. As the season progresses, however, leaching levels increase, especially in the wetter areas such as Mangwende. Lysimeter studies on sandy soils in Zimbabwe show that around 50kg/ha of nitrate can be lost in a wet year through leaching (Hagmann, 1994). Thus over the season the availability of nutrients varies dramatically, making the timing of planting and the addition of nutrients from fertilizers or other sources a key to success. In studies carried out near Mangwende, Grant (1967a) found that considerable losses occurred with the application of manure and most crops could not make use of any nutrient flush. The imperative of split applications of inputs to increase the synchronization of nutrient release and plant growth is clear.

Between years very different conditions apply. The coefficient of variation of annual rainfall in Chivi and Mangwende is respectively 37 per cent and 25 per cent. In the relatively recent past a cyclical pattern of rainfall across years was detected for southern Africa, with cycles of between 13 and 19 years observed (Tyson, 1986). Distinct wet and dry periods could therefore be detected. For example, most years during the 1970s had above average rainfall, whereas the 1980s were below average. Across the decades, then, different limiting factors operate. Thus in wetter periods, nutrients are limiting and significant additions of external inputs have to be sought. By contrast in the drier periods, water scarcity was particularly acute, and nutrient limitation rarely a problem except in low-lying areas with residual moisture. Different patterns of soil degradation can also be observed over time. During wetter periods, water-induced sheet or gully erosion is particularly important, with much erosion occurring as a result of a few key rainfall events; whereas in drier periods, soil dessication, compaction and wind erosion are more important degrading processes.

Understanding the changing nature of ecological dynamics is thus critical if we are to get to grips with the key questions central to soil-fertility management. For example, answers to such questions as: What are the limiting factors to plant growth? How does productivity potential vary over time and space? How are inputs applied most efficiently? all depend on an understanding of the complex interactions of ecosystem components and drivers. These in turn are affected by social, economic and political factors over time. It is to this history of soil-fertility change within Mangwende and Chivi to which we now turn.

A BRIEF HISTORY OF SOILS AND LANDSCAPES

From the 19th and into the early 20th century both areas were characterized by sparse populations making use of valley bottom wetlands (*dambos*) for rice and root crop cultivation. This was combined with limited shifting cultivation of grains (notably millet and sorghum) on the uplands. Such a system of intensive gardening using complex mounding systems to regulate water flows and enhance fertility levels relied on hoe cultivation and organized labour (Wilson, 1990). Thomas Leask, travelling in southern Zimbabwe in 1867, commented:

> *The hill was surrounded by rice gardens. These rice fields are in low swamp places and, the better to hold water, they are under ridges.* (Wallis, 1954, p114)

Large, extended households were the basic unit of production, with 'big men' being able to mobilize tribute labour for cultivation. Livestock, it appears, were limited at this time, in part due to Ndebele raiding (notably in the south) and particularly following the *rinderpest* pandemic of 1897.

However, around the turn of the century, and following colonization, agricultural practices began to change dramatically. The arrival of the plough, first picked up from settler farmers and missionaries, was a highly significant event in this period. The native commissioner (NC) for Chivi commented in 1910 that:

> *Natives are gradually acquiring a few more ploughs, but the movement is as yet slow. No wagons or carts have been acquired... the teachers and some of their relatives, on the mission farm 'Chibi', use ploughs and cultivate in European style, and in this way set a good example and teach the natives living in the vicinity.* (NC Chivi, 1910)[2]

By the 1920s, the opportunities for more extensive cultivation of dryland areas had increased, with plough use expanding in both study areas. In 1923 only 30 ploughs were recorded in Mangwende, but this number rose exponentially over the following decade, such that by 1932, 1400 were recorded in the district (NC annual reports). A similar pattern, although involving earlier adoption, was recorded in Chivi, where plough numbers increased from 18 in 1902 to over 5000 in 1939. A rise in cattle numbers in both districts (Figure 4.3) meant that draft power was increasingly available and an expansion into the land frontier was possible. By the mid–1920s the NC for Chivi noted: '... ploughing with oxen is considerably on the increase, one trader alone sold 200 ploughs' (NC Chivi, 1926). Ploughs became important symbols of wealth and progress, and became an important part of marriage negotiations:

> *It is difficult to estimate the number of ploughs brought into the District, as they come in from all sides. It is a not uncommon sight, however, to see natives wheeling them along the road, when out on patrol. This District, with others, shares the custom for a bride to refuse a suitor until he has purchased a plough.* (NC Chivi, 1928)

The creation of large arable fields was a labour-intensive process, involving clearing thick bush, often fending off wild animals in the process. Cultivation was an opportunistic affair with large areas planted, but only low yields expected. Ploughing was rudimentary and, at least in the first years, no additional fertilization was required. In 1922, the NC for Chivi complained:

> *Nothing is done to improve the light sandy soil predominating in this district although hundreds of tons of kraal manure are available and continual*

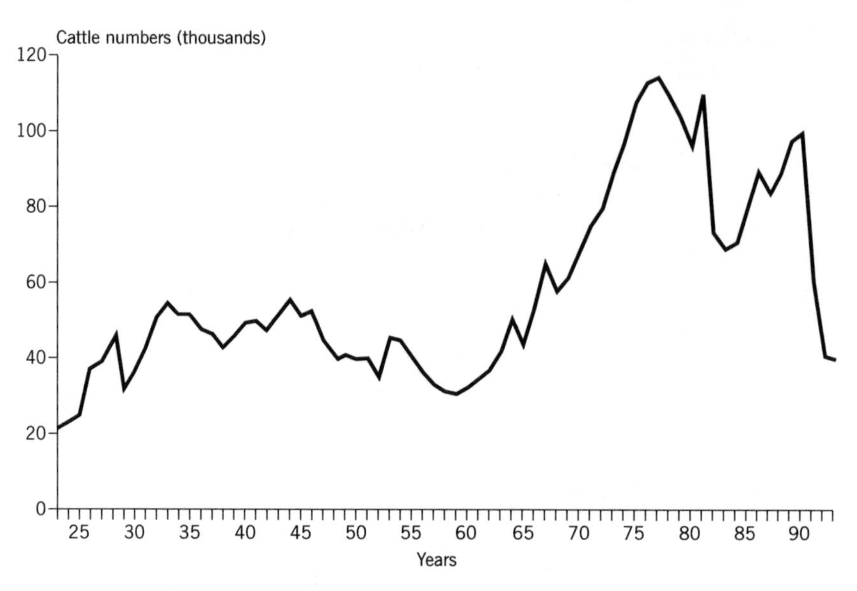

Figure 4.3a *Cattle populations in Chivi, 1923–1995*

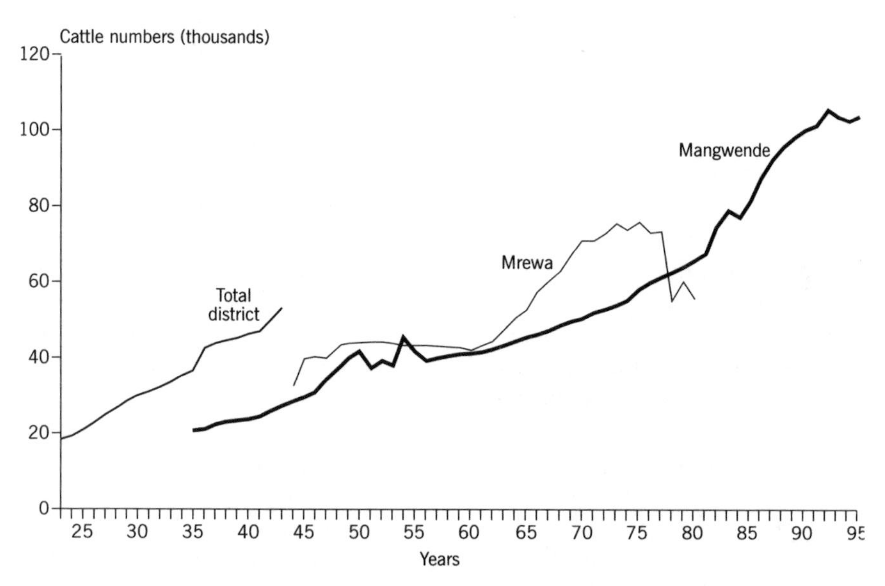

Figure 4.3b *Cattle populations in Mangwende, 1923–1995*

> *advice is given on this subject.* (NC Chivi, 1922)

Native commissioners' reports are replete with commentaries about the problems of inappropriate ploughing. For example, the NC for Chivi wrote in 1923 about the problems with plough-based land extensification:

> *Approximately 1300 ploughs are in use in the district, but it is questionable if there is any distinct advantage – except from a labour saving point*

> *of view. Ploughing is carried out in such an indifferent manner, that only a depth of a few inches is obtained. A large acreage is certainly cultivated, but usually more than the natives can cultivate properly with the result that the yield per acre is less than they obtained from the soil which was hand tilled.* (NC Chivi, 1923)

In Mangwende, the NC viewed the extensification process as highly damaging:

> *Another point against the plough is that some of the best land is on steep hill sides where only the hoe can be used, but having a plough the native cultivator picks out the poorer land on the flat and ploughs more of it to compensate for the quality.* (NC Mrewa, 1927)

When railing against the dangers of poor plough cultivation, the NC of Mrewa recalled the advantages of skilful hoe cultivation of the past:

> *With few exceptions the ploughing was done very carelessly. The land is not properly harrowed or drained with the result that a large part of it gets waterlogged and is useless. The native system of ridging their gardens showed to advantage in the heavy wet season we had last year. Newly ridged lands gave good returns, whereas crops planted on old grounds, where ridges had been washed out, gave very poor crops.* (NC Mrewa, 1923)

> *To see the lands alongside each other, on the same granite soil… is an object lesson. In the one done with the hoe, the contour of the ground has been studied, and the ridges made with a view to draining the lands from heavy floods but still to prevent erosion; the crop stands high and dry and in the heaviest rains and gets all the benefits of the humus turned under between the sods which form the ridges. The other is a scarified piece of ground…* (NC Mrewa, 1929)

At this time therefore, two forms of agriculture existed side by side: intensive hoe-based agriculture on the hillsides and in the valley bottoms, and extensive plough-based systems on the flat. This had major implications for soil-fertility management practice. In the extensive systems, very little fertilization was practised, with manure only being used on limited garden patches close to homes. For example, for Mangwende and Chivi:

> *Very few natives will use kraal manure to fertilize their lands, except in small patches near their kraals where they grow tobacco and sweet potato. Their objection is the strong growth of weeds which follows the manure and which causes them so much extra work.* (NC Mrewa, 1927)

> *I made every endeavour to get the natives to manure their lands this season, and at some kraals I have superintended the manuring of half an acre to an acre of land and hope to be able to demonstrate the advantage to be gained.*

> *My efforts have not, I must admit, been attended with much enthusiasm on the part of the natives, the complaints being that it entails considerable labour in carrying manure in small baskets from the cattle kraal and that it will mean a lot of extra work in weeding.* (NC Chivi, 1924)

By contrast, under the hoe-based shifting system, regular ash fertilization was important on the poor sandy granite soils. The NC for Mangwende commented in 1925:

> *It may seem a sign of indolence that the lands are not stumped before they are hoed and ridged up, but a little investigation shows that this is not the case, it is the result of experience gained… in most cases the only ground available is light granite soil. The sub-soil is never brought up, only about 3 inches of the top soil which contains all the humus and plant food is turned over by the hoe to form ridges in the gardens and on these ridges of about 6 inches of humus the grain is planted. The tree stumps are retained to prevent the excessive washing away of the soil. The branches are piled around them and burnt to act as fertilizer.* (NC Mrewa, 1925)

By the late 1930s, then, a new patterning of the agricultural landscapes of both areas was becoming evident. This involved intensive agricultural production in the wetter areas, combined with a more extensive system of opportunistic rainfed farming based on plough cultivation on the previously unfarmed upland soils. The growing livestock populations made use of the hillside areas for grazing which were increasingly abandoned as settlement and farming sites. The extensive crop lands also became a new, important source of fodder during the dry season as livestock became more closely integrated into the agricultural system. The NC for Chivi, for instance, commented positively in 1930 on changes in the agriculture of the district:

> *Ploughing of native lands, for the most part, is not the half-hearted effort that is experienced in other centres; the soil is usually turned over in even furrows, at a good depth except in places where there are outcrops of rocks; very little of the land is not turned over. Of course, there are mediocre and badly ploughed areas, on the whole natives cultivate their lands evenly and painstakingly. No harrowing is done. All seed is sown by hand. As the spread of improved agriculture continues more modern labour-saving devices will be utilised.* (NC Chivi, 1930)

However, serious investment in soil management was limited at this time. Land was plentiful, soil erosion rare and soil-fertility relatively high in the newly-cleared lands. However, this was not to last, and adaptations to farming practice were gradually made. This was first evident in Mangwende where land pressures were greater and soils inherently poorer. After a few years of cultivation without inputs, yield declines were noted as the available nutrients were used up or leached away. The resulting poor granitic sands required additional inputs if yields were to be maintained or improved. In Mangwende the increasing interest in new varieties of maize and European vegetables from the 1930s provided an additional impetus for this type of investment. However, such innovations were not immediate, and were initially focused solely on small home garden plots. However, by 1941 the district commissioner (DC) commented in his report that 'Slowly, year by year, a tradition of good farming is being built up'.

Although DCs observed some uptake of manuring by farmers in Chivi at this time, this was very limited. Although a number of agricultural demonstrators had been posted in the district since 1927, they had had limited impact. Similar frustrations were reported from Mangwende:

> *Those natives using crop rotation methods are still very few and far between. Not withstanding the efforts of the Agricultural Demonstrators, the rate of progress in this is very slow... The LDO [Land Development Officer] has been trying to get people to make compost... Like many other agricultural improvements, however, the native shows very little keenness when it means a little additional labour.* (DC Mrewa, 1945)

It was not until the widespread implementation of land-use plans, implemented as part of the centralization policy and later the Native Land Husbandry Act of 1952, that farmers were effectively forced to take up soil management practices. Centralization involved the reorganization of land into blocks, separating the arable land from the grazing areas with a series of linear settlements ('lines'). The arable land was to be on the uplands and the grazing in the low-lying areas. Environmental legislation enacted in the early 1940s (the Natural Resources Act) banned stream-bank and wetland cultivation, and restricted the establishment of gardens. Although not uniformly implemented, this major effort at land reorganization across the communal areas had a huge impact. From this time onwards, the two-pronged agricultural system involving both wetland/garden cultivation and dryland farming was effectively ended through central fiat. Where these policies were implemented vigorously (and this certainly included Mangwende, although less so in southern Chivi) the agricultural landscape was fundamentally transformed. In 1948, the DC for Mangwende commented:

> *An attempt will be made to protect the whole of the arable area of Mangwende reserve together with the individual allocation of land. It is felt that until this is done, no real progress can be made with the proper development of the reserve.*

By 1952 he reflected on the success of the programme:

> *The African is now fully alive to the benefit of protecting his lands with contours and in many cases now the kraal heads and their followers ask for the lands to be pegged and there is no difficulty in finding communities willing to contour their lands.*

These changes were associated with changes in soil-fertility management practice. With access to arable land increasingly restricted to the poor soils of the uplands, it became increasingly imperative to fertilize the soils. By 1941, the DC for Chivi observed:

> *Manuring of lands is increasing and would be more general were transport vehicles more numerous ... in order to increase the weight of the attack on bad farming methods several more demonstrators could be usefully employed.* (DC Chivi, 1941)

By 1947, the DC for Mangwende wrote:

> *Comment should be made on the extent to which compost making is now practised, and the use of kraal manure has now become the rule rather than the exception.*

Human populations in the 'reserves' also continued to grow. This was in part due to displacement from areas which had been designated 'European land', but also from natural growth. In Mrewa district, the estimated human population was 23,639 in 1923, and by 1943 it had grown to 44,749. Today's population is over 150,000. In Chivi a similar pattern was observed, with the district population rising from 28,500 to 57,220 between 1930 and 1962, with today's population estimated at over 150,000. At the same time livestock populations also grew significantly (Figure 4.3). With more men going out to work in the mines, on the farms or in jobs in towns, there was an increased investment of remittance income in cattle. Although attempts were made during the 1950s to destock cattle due to perceived overgrazing, this had only a temporary effect on the overall pattern of population growth. As draft animals and providers of manure for crop fertilization, cattle became increasingly central to the viability of the agricultural system in the communal areas.

From the 1960s, then, a pattern of intensification could be observed as a response to increasing land scarcity, forced in part by the colonial government's racial land policy, but also through a 'Boserupian' process (see Boserup, 1965) of population-induced innovation and intensification. Due to the differing agro-ecological conditions, combined with the contrasting access to markets, remittance flows, credit support and other factors, the pattern of change in Mangwende and Chivi diverged quite considerably from this period.

In Mangwende, a hybrid maize and fertilizer revolution took off. By the 1970s, a combination of close integration with the Harare market, good transport links, relatively efficient input supply networks and extension support through farmer groups, supported both by government and church NGOs, had encouraged the widespread adoption of fertilizer and hybrid maize. In good years, outputs were impressive and the more successful farmers gained considerable incomes from sales. In parallel, an expansion of commercial vegetable gardening occurred and many farmers opened up new gardens in *dambos* and other wetter sites. In the latter part of the 1970s, the liberation war intensified and the commercialization of agriculture in the area was inhibited for a period; but after independence in 1980, maize and vegetable production increased again. In the early 1980s, the expansion of the extension service and the provision of credit through the Agricultural Finance Corporation encouraged a large expansion of output from communal areas such as Mangwende (Rohrbach, 1989). This pattern was of course not universal, and some farmers were unable to afford the credit terms and purchase sufficient fertilizer and seed to really boost yields substantially. Marketing restrictions imposed by the parastatal Grain Marketing Board also kept prices depressed, and so limited commercialization to some extent. However, overall significant gains were made over this period. This was reflected in a fundamental shift in soil-fertility management strategy. Nearly all farmers purchased some fertilizer on a regular basis and, although manure remained important, it was not essential for maintaining yields. The really successful (and richer) farmers were able to combine manure and inorganic fertilizer and achieve significant yields in their homefields of up to 7–8 tonnes/ha in reasonable rainfall years.

In Chivi, by contrast, agricultural transformation has been slower. While nearly all farmers by the late 1970s were growing hybrid maize, few used fertilizer. By independence only 4 per cent regularly used fertilizer and this only grew to 9 per cent in 1986, despite favourable credit terms offered by the Agricultural Finance Corporation (AFC) in the early 1980s (Collinson, 1982; ICRA, 1987; Rohrbach, 1988). Thus, in contrast to Mangwende, a hybrid maize–fertilizer 'green revolution' did not take off in this period. However, a significant feature of agriculture in Chivi during the 1960s and 1970s was the establishment of the homefield as a distinct component of the farm. It was here that a limited form of agricultural intensification took place. With the recovery of cattle populations during the 1960s following the end of destocking, many farmers had plentiful manure to invest in their homefields. In many cases, the good rainfall years of the 1970s combined with high cattle populations and large quantities of available manure to make this a time of plentiful production of homefield maize. For most, the outfields were left more to chance, although some inputs were applied if available. Here millets and sorghums dominated, with some cotton and groundnuts and, by the 1980s, increasingly sunflower (Scoones et al, 1996). However, these crops received relatively few inputs, certainly compared to the now almost universally-favoured crop, maize. This pattern of production was upset, however, in the 1980s due to the series of major droughts which struck Chivi particularly hard. Cattle mortalities in the 1982–1984 drought were around 75 per cent, and this major loss was compounded by further deaths in 1986–1987 when around 20 per cent were again lost. The loss of manure for fertilization was a severe blow, and meant that farmers had to rely on focusing their soil-fertility management efforts even more on the homefield. Such scarcities, in turn, encouraged innovations in manure management and application (see below), and a growing skill among farmers in Chivi in the efficient management of organic soil-fertility resources.

By the 1990s, the context for agriculture in the communal areas had changed again. The combination of a major drought in 1991–1992 and the implementation of the government's structural adjustment programme from

1991 had far-reaching impacts on the agricultural systems of Mangwende and Chivi. The drought resulted in major mortalities of livestock, particularly in Chivi where around 60 per cent of the cattle died. Following the drought only 32 per cent of the households in Chivi owned any cattle (and most only one or two) (Scoones et al, 1996), meaning that manure as a primary fertilization resource was now really only available to a few in any significant quantity. The liberalization of agricultural markets as part of the adjustment programme occurred at more or less the same time, and resulted in a radical shift from the previous state-controlled system. This had the effect of boosting the marketing opportunities and producer prices for those with surplus to sell, and so particularly benefited the successful maize farmers of Mangwende. However, the down-side was the withdrawal of subsidies for inputs. The massive price escalation of fertilizer, for instance, has put some poorer farmers off. Fertilizer price changes have a limited impact in Chivi where few people use it anyway, but in Mangwende fertilizer had become central to the maize farming economy. Increasing entry of private traders and distributors has had a beneficial effect on supply and price, but the lack of credit markets restricts input-purchasing options for some.

In both Chivi and Mangwende, the consequence of these changes has been an increasing investment in gardening activities. In Mangwende this is particularly an expansion of commercial vegetable gardening on *dambo* sites, which require limited fertilization and have sufficient moisture. Maize remains the key crop for the homefields, and this is again where all available fertility inputs are applied. The exception to this is in cases where particular cash crops are grown. Increasing engagement with contract farming has meant that fertilizers and other inputs have been supplied as part of a package for planting such crops as cotton or paprika in the outfield. In Chivi, there has been an increase in the focused application of available fertility inputs – including the limited manure, but also increasingly composts, leaf litter and other organic materials – on gardens in homefields or on river banks. With fewer animals available for draft power and limited areas being intensively cultivated, this has often meant a return to hoe cultivation based on ridging and mounding systems, where the incorporation and management of organic materials can be effectively pursued. While the outfields have not been abandoned they are increasingly left to chance, with limited attention paid to ploughing, weeding and other activities; if something is reaped, then farmers are happy, but such fields are not relied upon.

This brief review has shown how the agricultural landscapes of both Mangwende and Chivi have changed dramatically over the past century, transformed at various junctures by both locally-generated processes of change and changes imposed from outside. The emergence of a mixed farming system, dependent on the harvesting of biomass from the non-arable areas by livestock and its transformation into manure for crop fertilization, has been a central feature over this period. Over time this has meant the transport of nutrients, via livestock and their *kraals*, from one part of the landscape to concentrate them in others in order to produce food. The precise pattern of this movement has changed over time as land use has been reconfigured due to land-use planning interventions and population movement, but the general pattern has been central to the sustainability of the system. However, the limits to this form of mixed farming have become increasingly apparent. Expanding arable land and declining grazing areas have made livestock more vulnerable to drought shocks, particularly in the dryland areas where grass biomass suddenly collapses in drought and few other feed options are available.

The decline in numbers of cattle per unit of arable land has meant that less manure is available, requiring it to be used in a more focused and efficient way. Other inputs have become important – in Mangwende external inputs of mineral fertilizers have compensated to some degree, while in Chivi the harvesting of biomass from the non-arable areas continues, but with direct applications or composting providing an alternative to manuring. With decreased availability of fertility inputs into the arable areas, the concentration of nutrients has increased still further. Today, most inputs are channelled to very limited areas on a regular basis and are associated with a single crop – maize – with the result that nutrients continue to build up in such sites and are depleted to low levels elsewhere. The patterning of nutrients in the landscape is thus increasingly heterogeneous with such nutrient 'hotspots', created by human action, being absolutely vital for agricultural production.

How farmers manage soils and their fertility is thus conditioned by these broader historical shifts in agricultural landscapes and practices. Such contexts are the backdrop for the emergence of particular understandings of soils and their management. The following section now turns to this subject, exploring local knowledges about soils and fertility.

LOCAL UNDERSTANDINGS OF SOILS AND FERTILITY

Farmers encounter a range of different soil types within the varied landscapes of Mangwende and Chivi. Understanding the different characteristics of soils is critical for effective agricultural management. Different soils may require different fertility inputs, different strategies for replenishment and enrichment, different matches with crops or different tillage approaches. Local classifications of soils in both sites take account of these contrasts, identifying a range of soil types. Box 4.1 illustrates this range for the case of Gororo, Chivi. Soils are identified in relation to visible characteristics of colour or texture, and this is related to a range of other characteristics important for crop management.

Such local classifications overlap with scientific ones, although not in a direct one-to-one mapping. Often the scientific classifications are pitched at a broader scale, with a number of different types, locally identified as distinct, lumped into one category. For example, the standard soil classification for Zimbabwe (Thompson and Purves, 1981) identifies only three types for Mangwende and Chivi. However, standard chemical or physical measures of soils are able to distinguish the soil types identified by farmers (FSRU, 1993; Nyamapfene, 1991).

Farmers' classifications of soils tend to be more fluid than the fixed classification used by soil surveyors. This is because farmers recognize that soils

BOX 4.1 LOCAL CLASSIFICATION OF SOIL TYPES, GORORO, SOUTHERN CHIVI

Musheche – Sandy soil, generally of low fertility; poor water-holding capacity; easily workable; requires fertility inputs, especially organic materials; suitable for most crops if fertilized. Including:

- *Musheche muchena* – Sands which are white and powdery and particularly infertile; manure and inorganic fertilizers required; groundnuts and roundnuts preferred.
- *Musheche mutema* – Poor fertility sands; require considerable inputs of fertilizers for reasonable crop growth.
- *Musheche mutema* – Black sands; moderately fertile, but crop growth improves with fertilization; suitable for most crops.
- *Musheche mutsvuku* – Reddish sands; require fertilization; finger millet and maize grown.
- *Musheche mujekese* – Relatively fertile sands with small pebbles; organic inputs required; groundnuts or beans do well.
- *Musheche mukabvu* – Good quality sandy soil; manure additions mean good crop performance; groundnuts and maize grown.

Chidhaka – Black, sticky, rich soil; prone to capping; not easily worked; fertility inputs not required; maize and sorghum grown.

Ganga – Black, hard, fertile soils; maize and sorghum do well.

Doro – Low-lying wetland soil; suitable for maize and rice.

Chishava – Red soils with medium fertility; limited inputs required; suitable for all crops.

Churu – Anthill/termite soil.

Chimwara – Stony soils in low-lying positions with reasonable inherent fertility; inputs not required; sorghum and sunflowers grown.

Jekese – Soils with plenty of stones; suitable for groundnuts, bambara nuts.

change and indeed can be actively transformed through their own effort. Soil-fertility is therefore seen as a dynamic feature of soils. Identifying which parts of a farm landscape are more or less fertile is an important part of farmers' expertise. This may relate to identifiable soil types and patches, as many farms may have a number of different soil types contained within them. However, finer-grained assessment within soil types may also be necessary and other indicators of soil quality used. A range of trees and grasses are regularly used for this purpose. For example, the presence of certain tree species (eg *mutondo*, *musasa*, *mupfuti*, *mupvezha*) and grasses/weeds (eg *derere*, *damba*, *mbavare*, *shanje*) are all indicators of richer soils where maize may be planted with limited inputs (FSRU, 1993).

Soils in different states of transformation – both positive and negative – can be identified by farmers in different areas of their farms. Local terminologies distinguish, for example, virgin soils (*gombo*), where agriculture has not been practised and the land has not been cleared. These may be inherently 'strong' soils (*ivhu rakasimba*), with good structure, organic matter and clay contents; or weaker, poorer soils – 'soils which require food' (*ivhu rine chikafu*). Such poorer soils need not stay that way and may be enriched to create 'fat' soils (*ivhu rakakora*) through a process of investment of inputs over a period of time. Thus, wandering through any farm, a history of soil change can emerge through discussion with a farmer, where some soils are in the process of enrichment, and others in decline.

How to encourage improvements in soil quality and avoid soil degradation is often a subject of conversation among farmers. Farmers have many views about the benefits and disbenefits of different fertility inputs (see below). In the lower rainfall area of Chivi, the importance of manure and other organic materials is continuously stressed, not only for its fertilizing qualities, but because of its importance in improving soil structure. In Mangwende farmers concur, but also argue for the benefits of inorganic fertilizer in combination. The problems of soil degradation are widely recognized, and many farmers have active strategies for preventing soil erosion.

Although contour bunding was widely resisted when first imposed, most farmers argue for its benefits today and actively maintain their ridges. However, the standard extension recommendations may be adapted to fit the particular circumstances of the farm. For instance, trenches or pits may be dug in the ridge to encourage water percolation and harvesting, while drainage channels may be created to direct water to productive soils and prevent erosion (Chuma and Hagmann, 1997). Of course, it is the visible water erosion and gully prevention where most effort is invested, and probably less attention is paid to soil and nutrient losses through wind, sheet erosion, rill formation, leaching or volatilization.

NUTRIENT-BUDGET ANALYSIS

Because of this variability across both space and time, great caution must be applied in the interpretation of snapshot assessments based on a limited sampling base. Such results must therefore be set within a wider understanding of the spatial and temporal dynamics of the ecological system. In the following sections the results of the nutrient-budget study are presented. This was carried out over a single season (1995–1996) and was based on the detailed study of a limited set of farms. Even given these limitations, the study does reveal immense diversity and variation across sites, offering some interesting insights into the way nutrients are managed on-farm.

Detailed studies of nutrient inputs and outputs were carried out on nine farms in each site (see Chapter 1). These nine farms were selected purposively from a broader sample (N=50) which was divided by local informants into three resource groups, representing different socio-economic characteristics and so different soil-fertility management practices. Table 4.3 shows how the different resource groups are differentiated according to a range of different criteria.

Table 4.3 *Socio-economic characteristics*

	Mangwende			*Chivi*		
	Richer	*Medium*	*Poorer*	*Richer*	*Medium*	*Poorer*
Homefield (ha)	1.7	0.9	0.8	0.8	1.3	0.8
Outfield (ha)	2.4	1.8	1.0	3.6	1.9	2.0
Cattle (no)	9.9	5.1	0.0	7.5	2.7	1.4
Scotch cart (no)	0.8	0.5	0.1	0.8	0.6	0.5
Family size (no)	6.3	6.2	6.9	10.2	7.6	8.4

Within the selected farms, a number of different niches were identified through a resource-mapping exercise carried out with the farmers (see Figure 4.4). Within the farm, the distinction between the homefield and the outfield was emphasized in all maps, although some farmers with limited land areas had effectively no outfield and treated their whole land area as a homefield. Within these broad categories a range of different niches were identified. These were associated with different soil types, topographical variations and past uses. Different crops were usually planted in these different niches, so sampling of inputs and outputs within niches was reported in relation to crop type.

For each niche around ten sub-samples were taken to plough layer at the beginning of the season (November–December). These were then mixed and taken for analysis. Table 4.4 presents the data on available N and P for homefields and outfields, with standard errors reflecting the variation across niches within these fields.

The data show clearly the differences in the nutrient status of homefields and outfields, with areas closer to the homestead having higher fertility levels due to the continuous input of nutrient sources over time. The only exception to this pattern is the levels of available N found in the outfields of Mangwende. However, the mean figures are distorted by a few unusually high values, and the high degree of variation around the mean suggests that many niches show very low levels of N availability indeed.

Variations in soil-fertility status across socio-economic groups did not show a definite pattern in both sites. The middle resource group had the highest average soil-fertility status, although when the total area available with higher-fertility status soil is considered, the richer resource group had consistently higher land holdings where N and P levels were above 30ppm.

Table 4.4 *Chemical characterization of farm components (available N and P in ppm and Standard Error [SE])*

		Chivi			*Mangwende*		
		RG1[No 39]	*RG2 [20]*	*RG3 [25]*	*RG1 [14]*	*RG2 [11]*	*RG3 [9]*
Homefield	Avail N	38.5 (15.3)	45.0 (19.0)	28.9 (10.9)	26.2 (5.6)	29.1 (9.0)	25.4 (9.8)
	Avail P	17.0 (14.1)	29.9 (40.0)	11.1 (15.9)	35.6 (17.0)	42.2 (43.0)	31.6 (20.0)
	pH	5.3 (0.8)	5.9 (0.5)	5.1 (0.7)	5.2 (0.6)	5.3 (0.9)	5.4 (0.8)
Outfield	Avail N	31.6 (12.6)	36.5 (23.6)	–34.1 (18.0)	45.8 (32.0)	33.0 (29.0)	
	Avail P	13.7 (9.0)	18.2 (26.3)	–12.2 (4.9)	8.5 (5.0)	10.5 (6.5)	
	pH	75.8 (0.6)	5.3 (1.0)	–5.2 (0.6)	5.3 (0.5)	4.9 (0.3)	

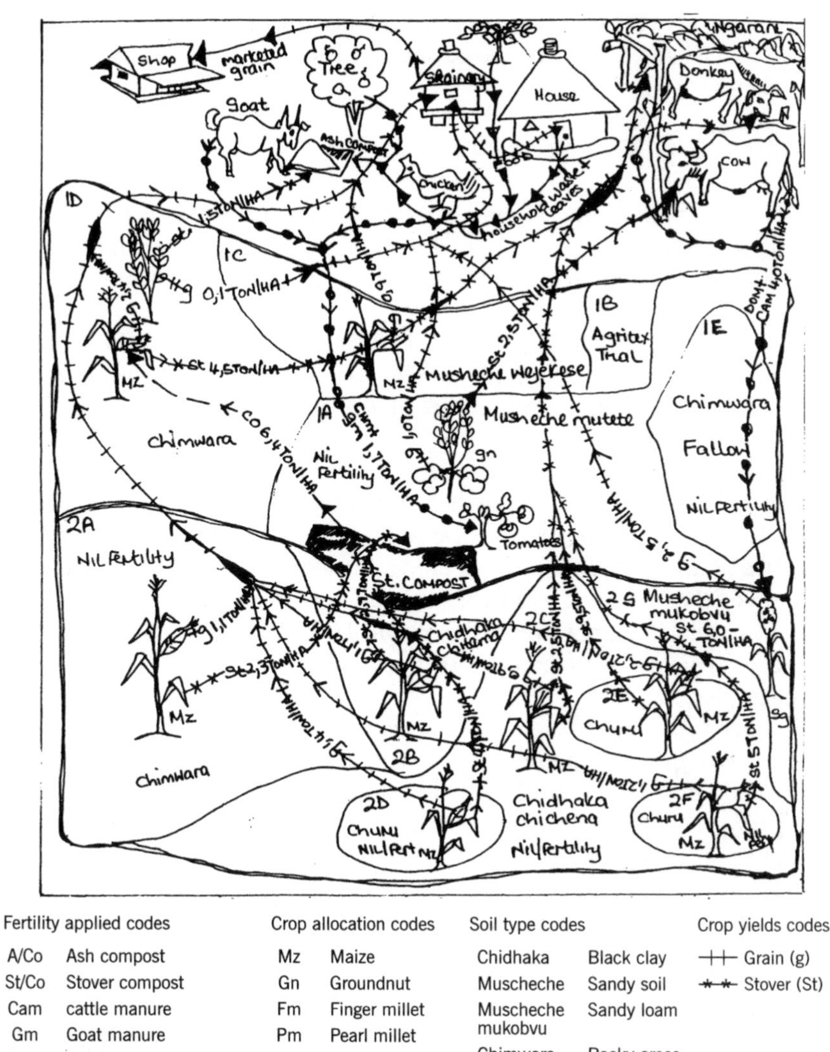

Fertility applied codes		Crop allocation codes		Soil type codes		Crop yields codes
A/Co	Ash compost	Mz	Maize	Chidhaka	Black clay	Grain (g)
St/Co	Stover compost	Gn	Groundnut	Muscheche	Sandy soil	Stover (St)
Cam	cattle manure	Fm	Finger millet	Muscheche mukobvu	Sandy loam	
Gm	Goat manure	Pm	Pearl millet	Chimwara	Rocky areas	
Chm	Chicken manure	Sg	Sorghum	Jekese	Gravel soil	
Dom	Donkey manure		Field boundary			
	Household waste/leaves		Niche boundary			
	Livestock manure					
	Compost					

Note: Field I has niches 1A to 1E; Field 2 has niches 2A to 2G. Niches recorded according to soil variability

Figure 4.4 *Resource-flow diagram drawn by a farmer from Chivi*

A partial-balance model was constructed for the 1995–1996 season where annual flows of both N and P between compartments within the farm homefield, outfields, livestock *kraal*, markets and homestead were recorded. This involved regular recording throughout the whole season of all inputs and outputs from each niche identified on the farm map and resource-flow diagram. Inputs from wet and dry deposition, biological nitrogen fixation and sedimentation were not calculated. Similarly leaching, volatilization and erosion outputs were also not assessed (see discussion of limitations of this approach in Chapter 1). Thus the balance analysis was based on visible flows only, where:

$$NUTBAL = IN1 + IN2 - OUT1 - OUT2$$

Where: $IN1$ = inorganic fertilizers; $IN2$ = organic fertilizers (including manure, household waste, compost, leaf litter etc); $OUT1$ = grain harvested and removed from the field; $OUT2$ = crop residues removed from the field.

A summary of the results is presented in Table 4.5. This shows the N and P balances for both homefields and outfields for each resource group across the two study sites. This represents an average (weighted by area) across all niches within these field types.

The results show how, for most fields and for most resource groups across both sites, N (and to a lesser extent P) are being depleted as outputs exceed inputs. However, this negative picture is not universal, since some homefields in particular are apparently accumulating nutrients. In order to get a deeper insight into these results, it is necessary to get behind the aggregate figures and look at what is happening within particular niches within fields, and how different farmers' management practices influence nutrient status at any particular place.

In both Chivi and Mangwende, maize, in both homefields and outfields, shows a quite different pattern to other crops. In Mangwende, maize fields show highly positive N and P balances in homefields, with outfields being marginally negative for N and more or less in balance for P. By contrast, niches associated with other crops in the same farmers' fields show highly negative balances for both N and P, with sunflower plots in particular showing high negative balances across all resource groups. A similar pattern is seen in Chivi, where maize fields show positive (or near-balance in one case) balances in both homefields and outfields and across resource groups. Again, other crops (groundnuts, pearl millet, bambara nuts, finger millet and sorghum) show largely negative balances, with groundnuts showing the greatest deficits. The only exception was one sorghum outfield which received a very large input of termitaria soil in the year of study which resulted in a strong positive balance. In other words, maize fields, particularly in homefield sites, receive special treatment, receiving the largest portion of inputs from both inorganic and organic sources. Other parts of the farm, by contrast, do not, with some crops such as millet or sunflower receiving practically no inputs at all.

Table 4.5 *Nitrogen and phosphorous balance (kg/ha) for different field types and resource groups in Mangwende and Chivi*

	Mangwende						*Chivi*					
	Richer		*Medium*		*Poorer*		*Richer*		*Medium*		*Poorer*	
	N	*P*	*N*	*P*	*N*	*P*	*N*	*P*	*N*	*P*	*N*	*P*
Homefield	50.8	16.5	–20.6	6.4	–6.0	0.6	–12.9	12.3	–64.9	–1.2	–28.3	–2.8
Outfield	–19.4	0.1	–7.8	–3.7	–22.7	–1.0	–39.9	0.1	16.5	29.2	–	–

MANAGING SOIL-FERTILITY: INPUT USE

The management of inputs – whether inorganic fertilizer, manure, compost, termitaria, household waste or leaf litter – is clearly critical to the final nutrient balance found in a particular site in a particular year. This involves complex trade-offs. If too much is applied, and little rainfall falls, crops may burn; if the input is spread too thinly, and rainfall comes in a heavy shower, much can be lost. In addition, different combinations of inputs may be appropriate for different soils and different crops. This section highlights some of the input management practices employed by different farmers.

Farmers, not surprisingly, hold strong views as to what is appropriate for their farm setting. Box 4.2 contains some quotes from interviews with farmers in Chivi on the relative pros and cons of inorganic fertilizers and manure, while Table 4.6 compiles information from across the sites derived from a series of such interviews on the benefits and disbenefits of a range of fertility inputs. Table 4.7, meanwhile, gives some more detail, showing how different inputs are focused on particular soil types and crops.

Table 4.8 shows what percentage of farmers apply which inputs to different parts of the farm, including gardens. A few clear patterns emerge. First, far more farmers apply fertilizer in Mangwende than do in Chivi. Second, in Chivi, where inorganic fertilizer is less of an option, a wider diversity of organic sources are used. Third, gardens and home fields are the primary destinations for such inputs.

Table 4.6 *Pros and cons of different fertility inputs from farmers' perspectives*

	Advantages	*Disadvantages*
Inorganic fertilizer	Crops mature earlier; strong green plants; good growth; yields high; easy to apply.	Crop burn if rains are poor; poor taste of crops; destroys the soil; expensive.
Manure	Big grains and cobs; soil structure and fertility improved; residual effects; mixes with compost/ termitaria.	Crop burn if rains are poor and applied alone; increases weeds; labour intensive production and application.
Household waste	Good for reclamation of degraded soils/sodic patches; promotes water retention; easiest and cheapest source for poor people.	Crop burn if rains are poor; small amounts produced.
Compost	Good yields result; soil structure and fertility improved; *striga* controlled; well-rotted compost does not burn; good for vegetables.	Crop burn if not well-rotted and rains poor; first-year effects may be poor; labour intensive to make.
Termitaria	Improves sandy soils; reduces waterlogging; improves yields, especially of groundnuts; reduces *striga*; long residual effects.	Crop burn if rains are poor; labour intensive; access to mounds may be limited.
Leaf litter	Soil structure and fertility improved; rehabilitates degraded/compacted soils.	Can promote *striga* infestation; collection areas may be far; access reliant on scotchcarts; labour intensive.

Source: Field interviews in Chivi and Mangwende

BOX 4.2 VIEWS ON INORGANIC FERTILIZERS AND MANURES FROM CHIVI COMMUNAL AREA

'Fertilizers can be dangerous especially in dry areas with little rainfall. Fertilizer can burn up crops to nothing if little rain is received. But I use fertilizer for quick action. The action can take only some hours, especially in the rainy season. I purchase fertilizer every year, since my sons and daughters give me money every month' (Mawarire, Chivi, 7.12.94).

'Fertilizers are good compared to any other input. This is so because fertilizers are instant products. A farmer can use fertilizer according to rainfall distribution. With more rainfall a farmer can use more, but with little rainfall, a farmer can use little or ammonium nitrate (AN) only. This is different to other inputs that are applied before rain falls. So the application of other inputs is dangerous because a farmer can apply without the knowledge of rainfall distribution' (Mr Muguta, Chivi, 16.2.95).

'Inorganic fertilizer is cheap to apply, as I can apply it alone. It also gives more nutrients to the soil. In good years, crops grow very fast and give good yields. Compound D is spread along ridges before planting. AN is applied using a teaspoonful on each plant. In order to increase the efficiency of use, it is best to apply on a manured field. Losses due to leaching can be reduced by this, as well as by cultivating regularly to avoid capping of the soil' (Mr Virukai, Takavarasha, 26.1.95).

'I apply manure to the poor sandy soils in the fields close to my homestead. Manure can work for more than three years without applying anything and producing better yields. In good seasons, manured areas are always producing good yield and good grain... Timing is needed on manure application. In a dry place like this, I put manure in small quantities. The yield can be reduced by poor manure application' (Benjamini, Chivi Central, 17.12.94).

'Although we face problems of rainfall distribution, manure can do better for yield production. Manure is the source of soil nutrients. When the soil is supplied with manure the yield will be of high grade. Whenever the soil is supplied with manure, the crops will be growing very fast... With manured places and enough rainfall, then it will mean good yield for the farmer. So you will find that successful farmers are farmers with manure to feed the field' (Chitando, Chivi Central, 7.2.95).

However, the data discussed so far do not show how input use is differentiated according to socio-economic resource group. In Mangwende the richer resource group farmers applied by far the largest amount of manure, as well as the largest amount of basal inorganic fertilizer. The middle resource group applied the most AN as top dressing. The poorest resource group, by contrast, applied no organic sources and only some inorganic fertilizer, focused on a smaller area.

The nutrient content of different inputs is immensely variable. Commercially-available fertilizers obviously have the highest nutrient content. For instance, ammonium nitrate (AN) top dressing contains 35 per cent nitrogen and compound D basal fertilizer has 8 per cent. By contrast, locally-available

Table 4.7 *Input application*

	Favoured crops	*Favoured soils*	*Residual effect*	*Timing*	*Coverage*
Inorganic fertilizer	Maize, cotton, sorghum	Sands	None	Before October buy basal compound D; before January buy AN top dressing.	A 50kg bag covers c.1 acre (for Compound D) and 1.5 acres (for AN). 50kg basal or AN approximately Z$120 (1995 prices).
Manure	Maize, vegetables, fruit trees	Sands	2–5 years	June–July dig out manure; August–September carry to field; November spread and plough under.	A *kraal* yields 15 cart loads plus, covering 0.5–0.6 hectares of heaps of c.90kg spaced 2–5m apart.
Household waste	Maize, pumpkins	Sands	1–2 years	All year: waste dumped in pit, including household waste, ash and farm yard sweepings; November: dig out and use in field.	Only about 20m squared per year covered, with c.2 wheelbarrow loads each 3–5m.
Compost	Maize, sorghum	Sands	1–2 years	January–April: collect grass, stover. October (following year): dig out; October/ November: spread and plough under.	Variable coverage, with 1–2 wheel-barrow loads each 3–4m.
Termitaria	Maize, groundnuts	Sands, gravels	3–5 years	June–July: dig termitaria; September–October: transport to field; November: spread and plough under.	One termitarium may yield 15 scotchcarts or more. 0.3–0.4ha can be covered, with heaps spaced 2–3m apart.
Leaf litter	Maize, groundnuts, vegetables	Sands	1–2 years	June–July: collect leaf litter; November: apply and plough under.	Variable coverage, with heaps of 1–2 wheelbarrow loads 3–5m apart.

fertility inputs have much lower nutrient contents. Manure, for instance, has nitrogen concentrations ranging from as low as 0.4 per cent to around 2 per cent depending on the sand content (which may be as high as 50 per cent) (Mugwira and Mukurumbira, 1984; Mugwira and Shumba, 1986). Termite soils also have relatively low nitrogen contents, although these are substantially higher than surrounding soils (Watson, 1977; Campbell et al, 1995). Crop residues of maize tend to have nitrogen contents of around 1 per cent, but groundnut residues have higher levels up to 2 per cent. Leaf litter has variable nutrient contents, with *miombo* vegetation containing around 1.4 per cent nitrogen (King and Campbell, 1994). High carbon-to-nitrogen ratios in some biomass resources

Table 4.8 *Percentage of farmers applying particular types of fertility input, 1995–1996 season*

	Mangwende			Chivi		
	Homefield	Outfield	Garden	Homefield	Outfield	Garden
Fertilizer	96	66	80	20	16	4
Manure	44	14	66	36	38	48
Ash	24	4	22	42	12	0
Compost	66	20	20	30	22	12
Termitaria	0	2	2	40	24	14
Leaf litter	10	4	24	6	4	26

(notably crop residues and leaf litter) mean that nitrogen release may be delayed, and decomposition may only take place over a number of years. Such high residual effect may be beneficial, although initial immobilization may result in the depression of crop yield in the first year following application. By contrast, well-rotted manure or compost may show rapid release, although this may be lost to effective crop growth due to leaching.

Thus soil-fertility management inputs are highly variable, dependent on the quality of the input, the socio-economic characteristics of the farmer and the placement practices used spatially. With inputs being focused in particular areas year on year, gradually soil-fertility levels change, transforming the soil-fertility pattern across the landscape to areas with high and low concentrations of nutrients in different patches. This variegated landscape created by human use is lain over the high level of spatial variability across slopes and between different geological formations.

One type of landscape niche that is particularly important in both areas is the low-lying wetland area (locally known as *dambos* or *vleis*). Such sites are important for garden production and year-round cultivation, even in the drier study area of Chivi. They are thus highly valued, and rights over such land are often heatedly contested (Scoones and Cousins, 1994). Box 4.3 explores two cases from Chivi and Mangwende where such wetland areas have become crucial to farming livelihoods. This niche requires a particular suite of soil-fertility management approaches.

EFFICIENCIES OF NUTRIENT USE

When a particular input is placed in a field, a variety of things can happen. Fertility applications do not automatically result in improved crop production: immobilization, leaching, washing or volatilization may occur and the benefits lost. Data from Zimbabwe show the potentials for absorption, particularly in clay soils; high levels of leaching in granitic sands which characterize most communal areas; volatilization from manure both spread on fields and in *kraals*, with major reductions in nutrient value and erosion of soil and nutrients, particularly through sheet erosion and rills. If the available nutrients in different fertility inputs are to be used efficiently, it is important to make sure

BOX 4.3 WETLANDS AND GARDENING

Case 1: Dambo farming – Jethro Bodo, Chivi

Jethro was born in 1950 and married in 1976. He now has 5 children. He started gardening on his own in 1968 and in 1974 he was given a dryland farm plot to farm on his own by his father. In the early years of his farming he grew the standard crops grown by others in the area – maize, millets, sorghum, groundnuts and sunflowers. But he found, by and large, they gave low yields and involved a lot of labour. When he returned from working as a driver in town, he decided to emphasize garden production, and gained access to a small patch of wetland (*dambo*) which was highly fertile and had year-round moisture. Today he has a highly productive garden including vegetables (tomatoes, sweet cabbage, rape), an orchard (mangoes, pawpaws, guavas, oranges, lemons, avocado, etc) and he also grows maize (especially for green consumption), sugar beans and sugar cane. In order to improve the productivity of the garden he has developed a complex water-harvesting system which channels water to his fields from the surrounding hills with stone-walling. The flow of water is slowed in the field by a series of contour bunds. Composting is an essential part of his soil-fertility management strategy. This is concentrated in the wetland fields where, because of sufficient water, he never experiences crop burning as he had when trying to improve the soil-fertility of his dryland fields. Trench compost is made in 1m-deep trenches near the fields where crop stover and grass is added during the dry season. The pit is covered with soil, and water is added to assist decomposition. In most years, the compost is ready for spreading on the field by the following February. Heap composts are made around the trees in the orchard in shallow basins where weeds, grass and other plant material is accumulated as a mulch. Finally, contour ridge compost pits are dug in the field and maize stover is thrown in. In combination with cattle manure from his 11 animals, the compost provides most of the fertility inputs. However, Jethro does make use of small amounts of inorganic fertilizer on his maize. This involves digging a small hole next to the maize plant after emergence and adding a 3-finger pinch of Compound D basal fertilizer. Later, he may add some AN top dressing using the same method. In 1995 he estimated he got a gross income of around Z$4000 from his small garden, far higher than he would have got when he focused solely on dryland crops.

Case 2: Vegetable gardening in Mangwende

Over the last decade there has been an explosion in gardening activities in the Mangwende area, as farmers – initially women and subsequently men – realized the potential of vegetable production on the low-lying wetland areas which criss-cross the area. The growth of the Harare market and the good and relatively cheap transport access has meant that the trade in vegetables has become exceptionally lucrative. While wetland cultivation remains officially illegal, government officials turn a blind eye, as the gardens being cultivated patently do not cause major risks of environmental degradation.

that nutrient release and plant uptake are well synchronized. Depending on the quality of the input and the type of soil, this will vary considerably. Some inputs may have slow release qualities, with long residual effects, others will provide a rapid pulsed release of nutrients to the plant. Thus it is not simply the balance of inputs and outputs which is of significance, but a wider concern with the efficiencies of nutrient use at various points (Chapter 1).

Making the most of fertility amendments means adjusting input levels in relation to water availability. In rainfed farming areas, where rainfall variability is high, as in Mangwende but especially in Chivi, this is essential. Many commentators have noted the absurdity of blanket recommendations under such conditions, and the potentially negative consequences for farmers' livelihoods if they follow them (Page and Chonyera, 1994). Fortunately, as we have seen, most do not, and adopt a form of response farming, where fertility inputs are adjusted depending on rainfall level and pattern, and applications of basal fertilizer or manure and top dressing split. Experiments in Zimbabwe have shown that, with such adjustments, up to 40 per cent increases in profits are realized over existing fertilizer recommendations over a 5-year period (Piha, 1993).

The management of water, through micro-irrigation and water harvesting systems (see Box 4.3), therefore becomes a critical part of improving input use efficiencies. With more stable levels of soil moisture, as in wetland plots or river bank garden sites, higher and more regular levels of fertility input can be applied, with a greater chance of building up soil-fertility over time. In more dryland areas, a more opportunistic, response-based system is more appropriate.

In all settings, attention to the quality and mix of inputs applied is key. For example, poorly-decomposed organic materials such as crop residues may in fact reduce available nutrients for plant growth for a period, whereas a well-decomposed compost or manure mix may provide a good quality source with slow release of nutrients over several seasons. Because of the great variation in the properties of different inputs, amendment strategies which involve using a combination of different sources may result in increased efficiency and reduced risks (see Chapter 1).

Increasing the level of organic matter in the soil, and the associated biological activity, is a key part of any strategy to increase input use efficiencies. Rotational systems, legume plantings, intercropping combinations, various types of agroforestry, and fallowing have all been advocated as potential solutions for the communal areas of Zimbabwe (Kumwenda et al, 1995; Hikwa and Mukurumbira, 1995; Mukurumbira, 1985). Indeed, since the 1930s variations of such interventions have been central to extension recommendations (Metelerkamp, 1988). But, in practice, the adoption of such strategies has been limited. While farmers are well aware of the benefits of rotation, legumes and fallowing, they rarely make use of them. Most rotations are limited and the most common crop sequence on over 800 plots in Chivi between 1988 and 1991 was maize, maize, maize (Scoones et al, 1996). While legumes, such as groundnuts and bambara nuts, are part of the cropping system, they are not necessarily planted systematically as part of a rotational system to improve soil-fertility. Similarly, fallowing is rarely used in such a way, as fields tend to lie fallow only when there is insufficient draft power or labour for planting. Experiments have also shown how such solutions may have technical limitations also. For example, on the very poor sands of the communal areas, extended fallowing has limited impact on soil-fertility build-up because of the low biomass production and limited biological activity (Grant, 1981). Legumes, equally, may not fix sufficient nitrogen without 'kick-start' applications of inorganic fertilizers under such conditions (Giller et al, 1997).

In practice, then, it is the careful timing and placement of available inputs which is key. Whether the intention is to slowly build up soil-fertility in a particular site, or whether a limited nutrient source is being used simply to maintain crop yields in an already poor soil, similar principles apply. But these are conditioned by a range of social, economic and institutional factors which influence different farmers' abilities to employ different soil-fertility practices. It is to these issues which we now turn.

PATHWAYS OF SOIL-FERTILITY CHANGE

Previous sections have highlighted the immense diversity in soil-fertility management strategies employed across the case study areas. What is the impact of such efforts over the longer term, both in terms of crop yields and the sustainability of the soil resource base? These are difficult questions to answer, but what limited time series data we have suggest the answer is far from simple.

The broad pattern of maize yields for Chivi and Mangwende since 1980 (as shown by Agritex data) is one of high levels of variation around a fairly low mean (of 1500kg/ha for Mangwende and 470kg/ha for Chivi), following patterns of rainfall increase and decrease. From this limited time series data no long-term declines in yield levels are observed. Indeed data for longer periods indicates that average maize yields have increased (due in large part to variety improvement). It appears, then, that fertility inputs to maize remain sufficient to maintain yield levels, and that yield variation is more associated with rainfall levels than anything else.[3] Although data is more scanty for other crops, this appears to be broadly the case for the two sites.

But is such production sustainable in the longer term? Nutrient budget data presented earlier (particularly for outfields) suggest considerable nutrient extraction. Since, in most cases, cultivation has occurred at a site for decades, one would expect a decline in yield levels given the relatively low level of nutrient stocks evident in the poor sandy soils which dominate both sites. So what is happening? Is such a yield collapse around the corner, or is there a more complex story involved?

Earlier discussions have emphasized the great diversity of soil management practices employed by farmers, and the different ways it is possible, even with very limited fertility resources, to increase efficiencies of use. It is to insights from this sort of data, rather than the generalized yield trends or nutrient balances, that we must turn if a more nuanced and differentiated picture of soil-fertility change is to be gained.

A more detailed examination of the case study farms studied for the nutrient balance work highlights how, often concealed beneath the aggregated data, a number of different pathways of change are evident. Four broad scenarios can be identified.

1 Fertility build-up in the homefield, linked to intensive maize production through the regular supply of organic inputs over long periods, complemented with focused inorganic fertilizer application. This results in rising yield levels over time and a build-up of a significant nutrient stock, based on organic matter in the soil.
2 Fertility maintenance for ensuring yields. This strategy is again focused largely on maize, but this time in outfield areas. Sufficient inputs of inorganic fertilizer or manure are applied to ensure a reasonable yield. This results in stable but variable yield levels (depending on rainfall), but no build-up of soil nutrient capital over time.
3 Fertility loss, but yields maintained. A similar strategy to 2 is employed, but lower levels of inputs are required because existing stocks are high. Such sites might include clay soil plots, areas which were previously settlement or *kraal* sites, or fields which are part of longer term rotations where residual soil-fertility can be made use of in subsequent years. Yields thus remain stable (although variable), but nutrient balances may be highly negative over the short and medium term, depending on stock levels.
4 Low-level soil-fertility and opportunistic cropping. In some areas, soil-fertility has declined to a low level, resulting in a minimum potential yield. No additional inputs are applied and expectations of significant yields are minimal. Crops such as sunflower and pearl millet may be planted opportunistically in the hope of some yield if rainfall is reasonable. In such sites nutrient balances continue to decline (and may in time cross critical thresholds of low organic matter), and yields are always low.

Across the study sites, different pathways of change across these four types can be seen in different fields of different farmers (sometimes with different pathways existing alongside each other on the same farm). Box 4.4 provides some case examples.

Most farmers attempt to pursue Pathway 1, at least on a portion of their farm. This is usually on the homefield where maize is planted and the large majority of inputs are applied. Depending on access to resources, this may be achieved in different ways, either through intensive manuring (mostly farmers with large cattle herds) or through a more labour-intensive route, with composting and the application of leaf litter and household waste (mostly poorer farmers). In the rest of the farm, different sites may be associated with Pathways 2 to 4, again depending on available resources. For some, with access to inorganic fertilizer and/or manure, Pathway 2 is feasible in the whole outfield, although the number of farmers who are able to apply sufficient inputs across the whole area is limited. In most cases a patchwork of different pathways is evident, with some outfield areas with some yield potential but low fertility receiving inputs not applied in the homefield (Pathway 2); other areas relying on existing stocks (Pathway 3); while others receive nothing (Pathway 4).

BOX 4.4 PROFILES OF CASE STUDY FARMERS ACCORDING TO FOUR SOIL-FERTILITY CHANGE SCENARIOS

Scenario	*Mangwende*	*Chivi*
1	Mr Chivengwa (RG1) was allocated a virgin piece of land in 1979. He has improved the fertility of this area through the application of manure. Today he owns 14 cattle which produce 60 carts of manure from the *kraal*, bulked up by the addition of grass to the *kraal*. The manure is applied to the homefield, together with household waste. Top dressing but no basal fertilizer is added here. This means that this field is highly productive and produces a lot of maize. The outfields also produce adequate yield and annually around 12 bags of Compound D and 6 of AN are added to the maize plots.	Murezu (RG1) has nine contours which were allocated to him in 1968. These had been previously been farmed by the Dzigagwi family and since been abandoned. Despite people mocking him that he would only 'harvest hunger' from these fields, he has now improved the poor sands such that reasonable yields are realized. From 1970 to 1993 this was based on the regular application of cattle manure. Today with no cattle following the 1991–1992 drought, he uses a combination of termitaria and donkey manure to maintain the built-up fertility. Despite not having cattle, Gwese and his family (RG3) have managed to improve their fields' fertility considerably since their arrival in the area in 1981. This has been through a combination of termitaria and compost applied to the poor sands of the 'exhausted' outfield. Today, it is highly productive. The home field was virgin land in 1981, but they have managed to maintain its fertility through careful management.
2	Mrs Kagoro (RG2) has used cattle manure on her field since 1960. In the past, large quantities used to be transported there. Today, only sufficent to maintain the fertility is possible due to reduced cattle numbers. Currently the herd stands at only 7. Since the 1970s, manure has been supplemented with fertilizer on the maize fields.	Murembu (RG2) has 6 cattle and 5 goats and these are important sources of manure for the fertilization of his 11 contours of sandy soil. These inputs, applied in rotation together with certain crops, have been sufficient over time to maintain yields. When particular patches of poor soils are discovered, then additional inputs such as termitaria or compost may be applied.
3	Mrs Mhuka (RG3) settled in the area in 1975. The fields are all sands but of reasonable fertility. In the past the family had cattle and manure was regularly applied to the homefield. This area is quite fertile and it is where most farming effort is concentrated. This field is always winter ploughed, is planted with maize and receives all the manure and most of the inorganic fertilizer. The outfields receive some fertilizer, but only on patches where problems are evident.	Tengemhuri (RG1) owns a large field which, although sandy, has quite a good level of fertility. Before the drought manure was applied to the field regularly, but this practice has declined due to the lack of cattle. Fertilizer is only used if handed out by the government.

4	Dance Mapfaka (RG3) has a small field which he was allocated in 1987 after moving from nearby. Low-fertility sandy soils dominate the area, although there are a few patches of black soils. He has no livestock, cannot afford fertilizer and must rely on the application of ash and stover composts. He manages to get some additional fertility inputs to the field by placing stover in particularly infertile spots with the aim of attracting others' cattle in the dry season. This results in quite a considerable additional input of manure.

The data discussed earlier suggest that outfield maize yields are being maintained at a level above a low-level equilibrium, but are not substantially increasing, implying that Pathways 2 and 3 are especially common. The high levels of nutrient extraction recorded by the nutrient budget analyses suggests that Pathway 3, where stocks are being relied upon to maintain yields, is particularly important. This may be especially so for Chivi, where the application of inorganic fertilizer is limited. In Mangwende, most outfield areas, notably for maize, receive some fertilizer, making it more likely that Pathway 2 is possible, although this may only be feasible for richer farmers. Clearly, over time, as nutrient stocks become depleted, Pathway 4 may be the result, and farmers may have to switch to other crops and adopt a more opportunistic strategy.

Clearly Pathways 1 and 2 are the most desirable, both in terms of yield returns and the sustainability of the resource base. But pursuing such strategies may not be possible across the whole farm and for all farmers because of limited access to fertility resources. This makes investing in ways of improving nutrient-use efficiency absolutely key. With limited inputs it is essential to ensure, through effective timing and placing of soil amendments, that yield levels are maximized for a given input level. However, some level of running down of resource stocks may be inevitable in certain circumstances. This may make sense if, in the longer term, such areas subsequently receive inputs, and Pathway 2 is then followed. Given the need for crop yields in the present, longer-term soil investments may be a low priority for some, making Pathways 3 and 4 sensible options.

Trade-offs between different soil-fertility management options and the consequences for pathways of change occur across farms, and a single-plot view is insufficient to capture the wider dynamic. For this reason our focus is on how farmers look at their whole farm, seeing how different components of the farm landscape change in different ways. In order to understand why different pathways emerge on different sites, and how these change over time, we need to look at the wider farming context and the range of socio-economic factors that impinge on the day-to-day management and decision-making of farmers. For it is these factors which influence the emergence of different pathways in different sites and on different farms. In the following section, therefore, we look at a series of biographical case studies of farmers, linking their changing social and economic fortunes to strategies of soil management on different parts of their farms.

SOIL-FERTILITY MANAGEMENT STRATEGIES: CASE STUDIES

As we have seen, depending on socio-economic conditions different soil-fertility management strategies may be used. Choices of inputs, and how much is applied when and where, are all affected by such conditions. In the following sections, three broadly-defined soil-fertility management strategies are identified, and explored through different case studies from Gororo, Chivi. Through the case studies, an attempt is made to locate an understanding of soil-fertility management within an assessment of broader livelihood contexts. Through such an analysis some different pathways of environmental and livelihood change can be identified (see below).

Poorer farmers who have always applied limited inputs

Marieni Museva is a divorcee in her early 40s. She has no livestock and has never used inorganic fertilizers or cattle manure. She farms two small contours and a patch of land around the homestead. The soil is sandy and poor yields are the norm if no additional fertility is added. Marien has been farming for 10 years using compost, ash and termitaria. She follows carefully-planned strategies to maintain fertility in each contour. In the first year termitaria is added where maize is planted; next she plants finger millet with compost; in the next year more termitaria is added when grounduts are planted; and in the final year of the rotation no fertility is added and maize is planted again. In the homestead patch maize is always planted, and ash and household waste have been added each year. Marien argues that compost, especially that which has been decomposed with the assistance of termites, is the best fertility source, especially for maize. Termite soils are also good as they can strengthen weak sandy soils.

Mrs Paswera is also in her early forties and is married to an older man who is now an invalid. Since the early the 1991–1992 drought they have no cattle, and must rely on other fertility sources. A combination of ash, compost and termite soil is now used. Household waste, including ash, is dug from the pit nearby the house in August and carried to the fields in September. This is then ploughed in when planting takes place. The same applies to grass compost which is made in the fields to save transporting labour. A series of heaps are created among the crops which are added to through the rainy season during weeding and the cutting of grass along field boundaries and contours. Towards the end of the rainy season, soil is put on top of the heaps and decomposition encouraged. By August–September the compost can be spread across the field. Termite soil is dug during the dry season around June or July. It is then transported to the fields and spread by hoes in August.

Medium wealth farmers who relied on manure, but are now diversifying

Since his arrival in the area in 1952, Mr Manganeni has used manure as the main source of fertility input. Up to 1983, he broadcast manure in rotation on all his maize fields, spreading heaps across the whole contour. During the 1982–1984 drought however, he lost many of his cattle and he had to change his strategy. This involved economizing on the use of manure by applying it along the furrow in maize fields, and extending the range of fertility inputs to include inorganic inputs, ash and termite soil. Termite soil was first applied in 1984 to both maize and groundnut plots, placing a wheelbarrow load about a metre apart and then ploughing it in following spreading. Ash was first used in the main field in 1990 and applied to a maize plot with good effect. Inorganic fertilizers were used following the drought of 1991–1992 when Mr Manganeni's cattle herd was depleted yet further. Both compound D and AN top dressing were applied to maize, making use of a single 50kg bag of each with the basal fertilizer applied in the furrow and the top dressing applied to the side of individual plants.

Mr Ngwenya is of Mozambican origin and was born in 1923. He worked for a long period in the sugar estates from which he retired in 1987. He has 4.5ha of land, made up of largely very poor sands, although a low-lying portion on one edge of the farm has richer soils and more water available. These were allocated to him in 1980 by the local headman. Before 1990 he used manure as the primary fertility source for his maize fields. Large heaps of three wheelbarrows each were placed on the fields around October. These were then ploughed in to avoid losses due to wind erosion and volatilization. In those areas he could not cover with manure he applied both basal and top dressing, purchased from his earnings. Following the drought of 1991–1992, he was forced to change this practice due to depleted livestock and a lack of cash income. He diversified his fertility sources to include compost and termitaria for the main fields and leaf litter for the garden. Water harvesting techniques which have captured water flowing to the lowland portion of his farm have encouraged vigorous grass growth along the drainage channels and in the low-lying areas. He cuts fresh grass and allows this to dry in the sun for a few days. This, he argues, speeds up decomposition. The compost pits are filled from January to May, and by October he has a well-decomposed compost ready for applying on the field. Compost is applied to maize by placing a handful in planting holes which are dug with hoes. Termitaria soil is applied in the same way as he does not have many termite mounds on his land, although a broadcast method is used for groundnuts.

Richer farmers who make use of manure and fertilizer, but are now using inputs more efficiently

Mr Chinguwa was born in 1935 and started farming in 1955. By 1960 he was a Master Farmer and was growing new varieties of maize, finger millet and groundnuts. He used to own more than 50 cattle, although now he only has 18 thanks to the devastating drought of 1991–1992. Before this drought, he was able to apply generous quantities of manure, with heaps spaced as close as 2m apart. Winter ploughing in August was carried out when manure was incorporated. Since 1991, he has experimented with ways of increasing the output of manure through the addition of leaf litter, termitaria, crop residues and grass. Between January and February he may collect up to 25 scotchcarts of leaf litter to add to the *kraal*. He collects this from certain trees, particularly fig trees (*Mushavi*), which he says have good quality litter. He has a separate pit for household waste, but this too is dug out in February and added to the *kraal*. He finds this mixed source of fertility input is of excellent quality and it reduces the risk of crop burn compared to any of the inputs used alone. Even this large amount of bulked manure is insufficient to fertilize all his fields, and he continues to use fertilizer on the rest of the area, with both basal and top dressing applied to maize, finger millet and sorghum, and gypsum applied to groundnuts. Basal fertilizer is applied in furrows, and limited amounts of top dressing added later depending on the moisture conditions.

The histories of soils in a particular area are thus intimately bound up with a complex interaction of ecological, social and economic processes over time. Such histories are not neat, linear and predictable patterns of change, as a range of events, often sudden and unexpected, impinge upon them. Thus the droughts of the 1980s and early 1990s have had, as discussed earlier, enormous impacts on the possibilities of soil-fertility management. Similarly, wider changes in economic policies, most notably the structural adjustment impacts of the 1990s, equally have had major ramifications. What happens at a particular site, and which of the variety of pathways of change elaborated earlier results, will be the consequence of the interaction of such events with social and economic patterns and processes at a local level. As the case studies illustrate, a diverse range of social relations are implicated. For example, labour relations within and between households may result in different opportunities for investing in intensive garden production, as social norms influencing gender relations and agricultural labour, or local institutions mediating access to group forms of labour, may all influence what type of soil management is employed at a particular site.

Understanding how different people gain access to the key resources necessary for successful soil-fertility management is therefore critical to an overall assessment of the likelihood of both positive and negative pathways of change. The next section, therefore, explores in turn the institutional processes by which farmers gain access to land, labour and transport, capital, credit and savings, and knowledge, skills and technologies.

SOCIO-ECONOMIC ISSUES: ACCESS, CONTROL AND MANAGEMENT OF FERTILTY RESOURCES

As the data presented above has shown, different farmers apply different types and levels of inputs depending on socio-economic conditions. Access to particular types of resources – such as land, labour, capital and knowledge –

influences soil-fertility management practice. Access to such resources is mediated by a range of institutional factors, including tenure regimes, local social networks, gender relations, socio-cultural and political rules and norms, extension systems and project interventions, market functioning and so on. This complex of factors defines the range of livelihood options open to different people and, in turn, the possible strategies for soil-fertility management. Linking these elements together – access to assets, institutional mediation and livelihood strategies – helps to define the range of possible livelihood and environmental management pathways which are open to different groups of people in different settings (see Chapter 1 and below).

Access to land

In Zimbabwe, land in the communal areas is allocated by local headmen (*sabhuku*) to all residents of the area. During the Native Land Husbandry Act attempts were made at consolidating land holdings and allocating what were deemed to be viable production units of 6–8 acres. Local headmen were allocated larger portions in recognition of their administrative role and the requirements of feeding visitors to the area. Attempts to maintain this system of land allocation and prevent the expansion of arable lands into designated grazing areas have continued to the present, but with variable effect. In Mangwende, for example, continued expansion of arable land has occurred in response to population pressure. During the liberation war in the 1970s, land administration broke down and a movement of 'freedom farming' was encouraged by the liberation fighters.

Following independence, attempts were made to relocate farmers in line with land-use plans constructed in the colonial period, but this had limited success. In the study area in southern Chivi, land pressures have not been as intense. Substantial in-migration occurred into the area from the 1950s, first as part of forced removals from farms designated for European use, and later as part of a process of rural migration in response to land shortage, misfortune or displacement due to conflict (especially in-migrants from Mozambique during the 1980s). Due to the relatively low population pressures, no major attempts at land reorganization were attempted in southern Chivi, so land allocation remained in the hands of local headmen associated with the autochthonous lineage groups with long residence in the area. Today, land access continues to depend on allocation by local lineage leaders (although during the 1980s and early 1990s, confusion of authority in this area arose with the conflict between so-called traditional leaders and newly-appointed village development committees). In addition, inheritance is another route by which sons gain land through the sub-division of their fathers' land.

However, usufruct rights are conditioned by various factors, including social connections to the lineage heads, gender relations and the emergence of a land market in some areas. The informal basis on which land allocation is made means that social relations with those in control are critically important. It is perhaps not surprising that land allocated near good water sources, such as *dambo* land, or land with rich, fertile soil, is almost invariably occupied by members of the main lineage families (Scoones and Cousins, 1994). Although this is in part a reflection of the length of residence in an area, new land tends to be allocated preferentially.

Formal access to land by women is limited, and has been since the 1950s when rights of land ownership were formally attributed to men. However, in practice husbands may often hand over a portion of the land to their wives, and when husbands are absent, women are the *de facto* owners of the land. Gardens outside the main field areas are perhaps the only agricultural land exclusively managed and controlled by women, although men are increasingly engaging in gardening and claiming such land for themselves (see below). Access to good-quality garden land is particularly contested, especially as environmental legislation restricts options for riverbank or *dambo* cultivation (Scoones and Cousins, 1994).

Although the buying and selling of land is notionally illegal in the communal lands, as all land is formally owned by the state, land transactions involving cash are increasingly common, particularly for higher-potential land in Mangwende. Such factors therefore make access to good-quality land particularly hard for in-migrants, younger people, women and the poor. This makes soil-fertility management for such farmers, with access only to poor quality soils and so limited nutrient stocks, especially important.

But access to arable and garden plots is not the only facet of land access which is important to soil-fertility management, as communal grazing land is an important nutrient source for agriculture through the transfer of nutrients through livestock manure or directly through biomass (Swift et al, 1989). Tenure issues are important here, with most grazing land held as common property. The degree to which such tenure regimes are effective in maintaining the quality of the resource is highly variable, however (Cousins, 1990). With increasing pressure on common grazing land, access to other land outside the communal areas is also important, with 'poach grazing' and harvesting of grass or leaf litter occuring in neighbouring resettlement areas or private commercial farms in both study sites. Again access to these resources is changing, as such areas become increasingly occupied and, with this, boundaries defended.

Access to labour and transport

The collection, management and application of soil-fertility inputs may involve considerable investments of labour. It is not only the management and transport of bulky biomass based resources that is labour-consuming though. The careful placement of small quantities of higher-quality fertility sources may also require considerable effort. For instance, the placement of small amounts of inorganic fertilizer at the base of individual emergent plants inevitably requires a lot of time. Access to sufficient labour is required for fertility-management strategies that either rely on bulk transport of freely available biomass resources, or the use of small quantities combined with the maximization of nutrient-use efficiencies.

The timing of labour inputs is also key as soil-fertility management tasks may compete with other activities. Many soil-fertility management tasks take place in the dry season when there is less competition for labour resources. However, some tasks, such as post-emergence application of fertility amendments, may compete with other key agricultural activities. Labour requirements may be reduced with access to transport. The availability of carts and wheelbarrows, for instance, greatly reduces the labour costs of manure, compost or leaf-litter distribution, allowing farmers to put such inputs in far-flung fields with relative ease. However, as Table 4.3 showed, not everyone has immediate access to such transport resources.

The mobilization of labour or transport depends critically on social relations both within and between households. Again, this has a gender dimension with men and women often taking on different tasks. For example, it is men who generally take carts to far-off woodlands for the collection of leaf litter and who travel to dig termite mounds in the communal grazing lands. Whereas women tend to be more associated with the more localized transport of fertility resources, such as from the cattle *kraal* to nearby fields and gardens, as well as the application of the fertility inputs to the fields. However, such sexual divisions of labour are not fixed or constant. When men are absent, for instance, women often take on other tasks.

Social relations among households may be important in gaining access to labour and transport. For example, cooperative work parties (*humwe*) may be held, with a number of people invited to share food or beer and dig a termite mound or spread manure in a field. Equally, carts and draft power may be shared, allowing those who do not own a cart or have access to draft animals to transport fertility materials. Such networks of sharing and cooperation are usually based on lineage clusters, where kinship ties link households together. But this is not the only form of sharing network, and links based on friendship or simply the neighbourhood may also be important (Scoones et al, 1996). Those outside such networks are most certainly at a disadvantage and must rely on charity or the market.

In both Chivi and Mangwende, markets in labour and draft- or cart-hire exist. During a survey carried out in 1996, the hire of labourers for digging out manure from a *kraal* cost Z$60, while scotchcart hire cost Z$150, or Z$10 per load. Digging of termitaria cost around Z$100, while three days' grass or leaf-litter collection cost around Z$35.

Access to capital, credit and savings

Access to cash income is essential for the purchase of fertilizers and livestock. As we have seen, fertilizers are particularly important in Mangwende, where relatively higher rainfall and poor soils make the application of inorganic fertilizer an essential component of many farmers' soil-fertility management strategies. As already discussed, this is less the case in Chivi, where the returns from fertilizer applications are more uncertain due to unpredictable rainfall. Livestock is important in both sites for the provision of manure, but as discussed earlier, the repeated droughts of the last decade have had serious impacts, particularly on cattle populations. Increasingly, many farmers are finding it hard to restock after drought, and the numbers held per household is in decline.

For some, regular sales of crops or livestock may provide sufficient income for the rebuilding of a herd or the purchase of fertilizer each year. Among the wealthier farmers in Mangwende, this pattern is the most common. But for others, cash must be sourced from elsewhere. For such farmers the link between off-farm income and agricultural production is essential.

In the Zimbabwe context, remittance income from relatives working outside the communal areas has historically been vital. From the 1930s, increasing numbers of men migrated to towns, newly-established mines or commercial farms for both seasonal and longer-term paid employment. Incomes from such jobs have been an essential route to building capital assets in the communal areas, most notably the establishment of a cattle herd, as well as investing in agricultural inputs and equipment. Today this still remains the case, but, as discussed above, the impacts of economic reform have meant declining wage levels in real terms and an increasing level of unemployment in the formal sector. Thus, for many, even if they have a job, surplus income for reinvestment in rural production is increasingly scarce.

Alternative ways of raising cash of course exist and a growing range of off-farm activities are observed in both study areas. These include trading, building work and craft activities. Many of these result in only marginal returns, however, and are insufficient for the type of large-scale investments required for purchasing cattle or buying sufficient fertilizer on a regular basis.[4] Instead, most such incomes go towards servicing regular needs, including the purchase of groceries and clothes, or the paying of school fees, funeral expenses or medical bills.

The functioning of markets – both for inputs and outputs – is particularly important for those aspects of the agricultural system reliant on the cash economy. With the liberalization of the agricultural marketing system following structural adjustment in 1991, major changes have occurred. In many respects this has been to the advantage of agricultural producers, particularly in areas with good transport connections. Thus for Mangwende farmers, new markets have opened up and problems with input supply have not been faced. However, in the remoter areas of the Chivi study area, this has not been the case, as traders supplying inputs or purchasing outputs have to add substantial premiums previously absent with the pan-territorial pricing system.

Credit options for agricultural production are currently limited, although in the past, state-supported agricultural credit was an option. As discussed earlier, the Agricultural Finance Corporation offered generous credit terms during the early 1980s for the purchase of fertilizer and hybrid maize seed. But due to very large default rates, especially in areas such as Chivi where drought is common, this facility has been withdrawn. In Mangwende formal credit options do exist, but this is limited by the requirements of collateral imposed by lenders. Despite demand, informal credit systems are not prevalent although, as everywhere, the lending and borrowing of money between friends and kin is a common occurrence. This, however, is seen as helping someone out of difficulty, rather than as a regular provision of credit for productive investment purposes.

Savings institutions offer another route by which money is accumulated. In both sites informal savings clubs exist, sometimes associated with a beer-drinking circle, sometimes as part of a church group. In Mangwende such clubs have been promoted over a long period by the Catholic church as part of group development schemes. Such clubs take a variety of forms, some involving a large, mixed group of people, others much smaller and exclusive to men or (more often) women. They vary in the degree to which they are used to raise money for agricultural inputs and other investments, however. Although the amount of money circulating goes up and down depending particularly on the level of harvest and the amount sold (but also on the vagaries of remittances sent by husbands, sons and daughters), the amounts tend to be relatively small. Each disbursement, for instance, is unlikely to be able to purchase any cattle, but goats or small amounts of fertilizer may be possible.

For many, the prospect of being without cattle for long periods is a very real one. In the past when cattle were abundant, loaning arrangements meant that farmers without cattle could look after them and gain the benefits of manure in exchange for herding and looking after the animals. While such arrangements still exist, they are on a much smaller scale, and available only to close family members or friends. For those outside such networks, gaining access to cattle presents a real challenge. Market-based arrangements for draft hiring have emerged, and, in some areas, a market for manure has been observed. For most, however, getting access to manure means attracting other people's cattle to your field during the dry season, perhaps by heaping stover or even cutting branches from nearby trees.

In most cases, access to both fertilizer and manure (through the purchase of livestock) is then critically dependent on getting access to cash. This is mediated through a range of institutions, most notably informal kin networks which support the flow of remittance income. Formal systems for accessing cash through credit arrangements, in particular, seem to be limited in scope, and those that exist may be impossible for many. This makes gaining access to key capital assets such as livestock, and the cash purchase of fertilizer, problematic.

Access to knowledge, skills and technologies

The various soil-fertility management practices described in previous sections require different types of knowledge, a wide range of skills and access to a variety of technologies. Innovations in soil-fertility management come from multiple sources. Some come directly from extension advice, others are picked up from contacts elsewhere – from other farmers, from relatives in town, from experiences of working on commercial farms, and so on. In most cases farmers adapt such innovations to their own setting, with the result that a wide range of management practices and adapted technologies are seen on the ground.

Extension advice on soils management is widely available in both study sites from the national agricultural extension service, Agritex, as well as NGOs and others operating in the area. Agritex advice tends to focus on a standard set of packages, many dating back to the colonial era. For example, the standard mixed-farm model is central to much advice, with extension urging farmers to adopt rotational cropping systems, combined with high levels of manure input. Advice on fertilizer applications is also available, but a generalized blanket recommendation is offered which is rarely applied. Extension agents, especially in the colonial era, have been very active in promoting soil conservation measures, particularly the building of contour bunds. This was compulsory in the past and remains a central part of the training of 'master farmers'.

But soil-fertility, according to many farmers, is not wholly within human control. Hard labour and intelligent application of technologies and management techniques must, many insist, be combined with appropriately conducted relations with the spirit world. Some people, for example, are possessed by particular spirits (*shavi*) for good farming success. With the assistance of such sprits, it is argued, soil productivity increases and crops grow better. Others may call on the assistance of traditional medicine persons (herbalists, *n'angas*). Particular medicines or spells (known variously as *divisi*, and *tsvera*) can, it is claimed, assist in the improvement of harvests on one's field. *Divisi* can also be used negatively to harm yields and reduce fertility of the soil in others' fields, although additional medicines can be sought which protect such intervention (Mukamuri and Murwira, 1995).

Although many claim not to believe in such means of altering farming success, pointing to more 'rationalist' explanations for differential soil-fertility and yield, most residents of the rural areas will avoid entering the fields of others lest they be accused of depositing medicines in the field. Across the communal areas, prayers are held in the churches for good rainfall and successful harvests, and rain-making ceremonies, often attended by the same people, call for the same results by making offerings to the ancestors.

Most of the time, farmers quite happily seek 'scientific' advice from extension workers, while simultaneously following 'traditional' or religious practices in relation to soil management. Sometimes, however, different knowledges come into conflict. A recent example has been the conflict between formal extension advice and that offered by Ambuya Juliana, a young female prophet who has had an intermittently wide following across a number of communal areas in the south of the country, including Chivi. She argues that mechanical conservation measures and inorganic fertilizers in particular run counter to the wishes of the ancestors, and that everyone should abandon their use. A range of current problems, including drought, declining crop yields, environmental degradation and other major calamities, can, she argues, be attributed to the displeasure of the ancestors with current agricultural practices, and particularly those 'modern' and inappropriate methods advocated by extension.

But, as already mentioned, standard extension packages are not the only source of information. And, indeed, as discussed in relation to the case of soil bunding above, such packages are almost invariably selectively adopted and adapted to suit local conditions. Links to the commercial farming sector have been an important channel of information flow. Ploughs, for instance, were first seen on commercial farms and subsequently made use of in the communal areas in the 1910s. Other tillage techniques, include no-till and minimum tillage approaches, have also been transferred from commercial farms and tried out in the communal areas.

Over time, farmers in both study sites have therefore accumulated experience from a diversity of sources. Knowledge about soil management is embedded in rural social relations and the everyday practice of soil-fertility management by farmers in the communal areas, conditioned by the power, control and the influence of different players. Sometimes changes in practice have been quite rapid, as in the adoption of the plough or particular new crops, like the contract farming crops which have become a recent feature of the farming system in both sites. Other transitions have been slower, with shifts occurring over longer periods as new practices become embedded and new skills are developed. For example, skills associated with commercial vegetable gardening are an important part of the livelihoods of some wetland farmers in Mangwende (see Box 4.3). These, of course, emerge from long traditions of hoe cultivation, mounding and wetland management dating back to the 19th century. However, with the growth in the commercial market for vegetables in the last decade or so, new gardening skills have been required, often derived from the experiences of migrants with small market gardens in urban areas.

With increasing incentives to increase the efficiency of use of limited available soil-fertility inputs, much informal experimentation among farmers has taken place in recent years with manure management, mulching, composting and fertilizer-application techniques. Much of this knowledge remains tacit, part of the everyday practice of what people do. However, more demonstrable technological innovations have emerged, such as the composting systems described in Box 4.5, which are observable and are widely copied and adapted.

Unfortunately, in most instances the standard research and extension system does not pick up on such innovations, and fails to link them to more formal networks for sharing and learning. This decreases the potential for spread of useful innovations, although informal networks of sharing between farmers do exist. Thus beer parties, church services, funerals and other gatherings become foci for information exchange and skill sharing. Visits to relatives and friends, work parties and other opportunities to actually see other farmers' fields also encourage the spread of innovations. With migrancy so much part of the livelihood system of the communal areas, the opportunities to travel and learn about soil management practices in other areas increases, with the result that such sharing is not just localized exchange.

However, as will be discussed at more length in Chapter 5, the encouragement of linkages between formal and informal research and innovation systems have been part of the new initiatives of the Farming Systems Research Unit of the Ministry of Agriculture, as well as a variety of NGOs operating in both sites. Through the coming together of groups, usually based on an extended kin-based cluster of households, farmers identify a range of soil-fertility management issues that they wish to tackle. Through simple plot-level experimentation and farm-level monitoring, a range of issues have been explored and a number of different technological and management practice innovations have been shared (see case 4, Box 4.5 for examples). A key element of such participatory research processes has been the building of confidence among farmers for the testing and sharing of ideas. The networks

BOX 4.5 EXPERIMENTING FARMERS IN CHIVI COMMUNAL AREA

Case 1: Mhazo

Following the drought of 1991–1992 when most of Mhazo's cattle died, he decided to experiment with intensive crop residue treatment and composting as an alternative to manure and fertilizers which had sustained cropping on his poor sandy soils since he started farming there in 1973. Since 1992, with the aid of hired labour, he has dug over 20 large pits in different parts of the farm. He has tested different depths and sitings for the pits, and concludes that the best is a 1m-deep pit built along the contour, often just behind the contour ridge. Into these pits he piles layers of maize and other crop stover and soil. After a year, the compost is well decomposed and ready to spread across the field. The advantage of placing the pits within the fields is, he says, the increased water infiltration that results. The consequence has been that the soil moisture of the field has improved, particularly immediately below the pit, meaning that he can apply even more organic material without the fear of 'burning' (Interview, 6.2.97).

Case 2: Marien

Marien lives close to Mhazo in Gororo area of Chivi, and had observed his experiments with compost pits in the field. Her problem, however, was that she neither had the labour nor the volume of stover to warrant such an investment. However as she has no livestock and has never used any inorganic fertilizer or manure on her fields, she has also experimented with a composting system. This involves digging holes (around 2m wide and 2m deep) in the large termite mounds which are scattered across her fields and putting a mixture of crop residues, grass and leaf litter inside. The hole is covered and the compost is then ready for the following season. By contrast, her other compost pit close to the home may take two years or more before effective decomposition occurs, depending on rainfall (Interview, 18.1.95).

Case 3: Mawarire

Mawarire is a relatively rich farmer. He has a large field of over ten contours and a herd of cattle which supplies manure. He is also able to afford some fertilizer as his son and daughter send remittances home. He thus has sufficient resources to make choices about how to manage the soils in his land. Most of his soils are poor sands, with some scattering of clay and sodic patches. Over the years he has analysed the responses of different soil patches in his farm to different manure applications and rotations. In order to test this, he creates a simple plot comparison in a particular soil type and compares manure treatments. As a result of this experimentation, he has adopted a careful rotation of manuring suited to different parts of the farm which allows each portion of his land to receive some manure at least every 6 years. He comments that 'depending on rainfall, the soil uses up the manure only at a slow rate'. He regards manure as the essential fertility input. This is complemented by inorganic fertilizer which he started using after independence, after having been shown how to use it by an extension worker. He has experimented with a variety of ways of applying it and with different crops, but concludes that a furrow application of basal fertilizer to maize gives the best returns. In recent years, with the decline in his cattle herd due to drought, Mawarire has been concerned about how to maintain his manuring strategy. He has therefore started experimenting with various types of manure treatment, involving adding different mixes of stover, grass and water to the *kraal* and to pits where manure is dug out. This has meant that he has increased the volume of production of manure, and, when ensuring good decomposition, the quality has been ensured too (Interview, 7.12.94).

Case 4: Farmer Research Groups

Across Chivi there are now over 20 farmers groups each working on different aspects of soil-fertility management in collaboration with the Farming Systems Research Unit of the Ministry of Agriculture. One of the earliest experiments designed was to explore ways of improving the quality of sodic soils. These are found in patches on many farmers' fields, and always result in negligible yields. The Museva farmer research group chose a series of treatments including adding river sand, leaf litter and ash. The research team suggested a lime/gypsum mix. With a control plot, the treatments were compared over a number of years from 1994, with the conclusion reached by farmers and researchers together that a river sand treatment was the best, and probably could be improved with some addition of organic matter such as leaf litter or a grass mulch. The results of these experiments were shared each year at the annual feedback meeting organized in Chivi. This provides the opportunity for the farmer groups to share their results with each other, and other interested people who attend, including agricultural extension officials, NGO workers and others. Other farmer groups in other parts of the area confirmed the results of the Museva group with their work on sodic soils, and now the information is widely known and shared, with farmers in other parts of Chivi testing out the river sand option.

among farmers within and between groups as part of wider feedback and discussion activities at a district level have also been important features of the process. This has, in turn, suggested new roles for the formal research and extension inputs – as a source of new ideas and innovations, as co-researchers in farmer-led experimentation, and as facilitators of wider networking and sharing among farmers, researchers and extension staff.

CONCLUSION

This chapter has shown how most policies on soils and land management have been framed by a particular style of research. This has largely been technical in orientation and has focused on a standard set of soil conservation technologies and largely blanket recommendations for fertilizer and manure inputs. These have been premised on the assumption that a desired end-point of such interventions is a mixed-farm model, based on the integration of crops and livestock. However, as we have seen, such policy proclamations and extension recommendations do not capture the dynamics and diversity of actual practice. While elements of the mainstream views are clearly justified, these are seen to be far from universal. Indeed, an understanding of soil management drawn from a different set of conceptual and methodological bases suggests some different ways of looking at the problem (see chapter 1). A perspective based on historical analyses of environmental change, and one that integrates biophysical and socio-economic understandings, highlights how the status of the soil resource (in terms of, say, erosion levels or nutrient budgets) is the product of farmers' everyday soil-management practices over time. These are, in turn, influenced by a range of institutional factors that affect access to resources such as land, labour, transport, credit, skills, knowledge and so on.

It is this interplay between the social and economic sphere and the natural environment that results in the diversity seen on the ground, and the range of pathways of change which different farmers pursue over time. Some pathways result in increasing levels of farm productivity along with improvements in levels of soil nutrients, while others show different trends. Across the Zimbabwe study areas, some pathways are reliant on keeping a steady, sometimes increasing, supply of inputs. For some, particularly in Mangwende, this means ensuring a regular supply of inorganic fertilizer; for others, particularly in Chivi, this involves manuring. Such strategies are reliant on different institutions, with markets and supply chains being key for fertilizer inputs, while institutions governing cattle ownership or sharing and loaning are important for manure inputs. For those without access to the cash to purchase fertilizer or the cattle to supply manure, other pathways are evident. For some this involves increasing the efficiency of use of limited available inputs by investing labour, knowledge and skills in complex soil-amendment practices so that timing and placement ensures optimal uptake. Not all pathways show positive increases in productivity and sustainability. In some areas, yields are declining along with soil-fertility as the conditions for investment in soil improvement are absent. In others sites, soil nutrient balances may continue to be highly negative, while yields do not decline as the nutrient capital in the soil is being made use of.

Unpacking what factors result in which sets of trends in particular places is the key to any analysis which takes diversity and dynamics in soil-management practice seriously. The result is an analysis which questions many of the assumptions prevalent in mainstream policy and extension recommendations, and which argues for an approach to supporting soil management by farmers which is more attuned to the range of circumstances which are found on the ground.

Chapter 5

Participatory Approaches to Integrated Soil-Fertility Management*

Toon Defoer and Ian Scoones

Introduction

The case studies presented in the previous chapters have shown how soil-fertility management by small-scale African farmers is complex, involving multiple sources, processes and flows of fertility. There is great variability in how farming is organized and managed, determined by a wide range of socio-economic conditions and differences in access to resources. In addition, there exists considerable variation in the agro-ecological conditions under which farmers operate. What is practical and profitable for farmers depends on the prevailing production circumstances. Under these conditions, it is difficult for research and extension to arrive at tailor-made recommendations. Management of soil-fertility in such settings clearly requires approaches that are responsive to diversity in agro-ecological conditions and socio-economic circumstances. Conventional approaches to soil-fertility management often start and finish with technological questions and answers, neglecting the broader systems implications and socio-economic and institutional conditions determining farmers' abilities to manage their soils.

This chapter draws on field experiences in Mali, Ethiopia and Zimbabwe where a participatory and learning approach to investigating and acting upon soil-management issues has been developed and tested. Such an approach tries to go beyond the conventional framing of the soil-management question in purely technical terms, and draws on a variety of other perspectives more rooted in farmers' own understanding of resource management and system dynamics. Rather than proceeding directly to interventions based on scientists' recommendations, more adaptive learning through experimenting, monitoring and evaluating of alternative options for improvement are explored. Managing the widest variety of possible sources of fertility, both on-farm and off-farm, in the most efficient way is known as integrated soil-fertility management (ISFM) (Defoer et al, 1999a). ISFM not only deals with nutrients, but also with other elements of the soil that contribute to sustainable plant growth, including physical elements of the soil such as texture and structure and (micro-) biological elements. In addition, ISFM considers not only the technical aspects of soil-fertility. In ISFM, soil-fertility management is viewed within a systems context, with attention paid to the prevailing socio-economic conditions, natural resources, and institutional and policy contexts. ISFM does not intend to develop so-called technological messages to be transferred by extension to the farmers. ISFM instead aims at making the best use of locally available resources, based as much as possible on local knowledge and decision-making, including understanding that stems from research (Röling and Wagemakers, 1997). ISFM then promotes a wide variety of options that are further tried out and eventually fine-tuned and adapted by farmers under local conditions. ISFM is therefore, by definition, highly location-specific and subject to changes over time.

In order to promote ISFM effectively, there is a need for approaches that allow close interaction between farmers, researchers and extension workers. Since soil-fertility issues are highly complex and knowledge-based, the ISFM approach necessarily deals with the generation and management of information, and the facilitation of learning. The aim is to assist farmers in improving soil-fertility management through an approach to analysing the current situation, followed by step-wise planning, experimenting, and evaluating alternative soil-fertility management practices. Such an approach provides an alternative to the linear, transfer-of-technology mode so characteristic of conventional research and development systems, offering a more interactive way of analysing complex and diverse agro-ecosystems and a more participatory approach to involving multiple stakeholders in decisions about technology options for sustainable agriculture.

This chapter investigates how the three country teams have dealt with the methodological challenges of promoting integrated soil-fertility management. First, a brief review is given of how agro-ecosystems dynamics and diversity can be handled. Then the principles of farmers and scientists working together are outlined, before embarking on the teams' experiences in collaborative learning. Evidence is given of the need to adopt a process approach in order to successfully deal with integrated soil-fertility management. Setting up and developing such process approaches has been the focus of our work over the last few years, the experience of which is discussed in the following sections. The chapter concludes with reflections on the new challenges involved in adopting participatory learning and action research for integrated soil-fertility management.

AGRO-ECOSYSTEM COMPLEXITY AND DIVERSITY

Conventional agricultural science is reductionist by nature and generally defines problems by abstracting them from their natural situation. Of course there is no problem with abstraction and simplification for modelling or experimental purposes, if such abstraction relates to some key aspects of a farm system. But how do modellers or experimental scientists who rarely visit a real farmer's field know what is key and what is important? Too often assumptions are made about scale, landscape position, the relative importance of particular resource flows or pools, or the dynamics of the system which, in turn, influence the nature of the analysis and so the results (see Chapter 1).

For understanding real-life complexity, conventional experimentation, focused on the treatment comparision of a limited number of variables under highly controlled conditions, may be of less use than methods focused on monitoring changes in a more holistic fashion, with temporal and spatial variability centrally part of the design.[1] Understanding the spatial and temporal aspects of environmental change often requires looking at extremes. Very often extreme events are key to understanding fundamental transitions and shifts in dynamics. Equally, spatial extremes may be nodes of change and potential nuclei of innovation. Such a perspective in many ways runs counter to the conventional approach to experimental design and the basic premises of normal statistics. To avoid the 'tyranny of the average', alternative survey and experimental approaches are required which explore the ends of a population distribution and examine 'outliers' as important case studies from which key lessons can be learned.

Asking the relevant questions is of course an important part of good science. And good adaptive science must ask appropriate questions in relation to the likely users of its outcomes. This is almost so obvious that it should not need saying. But sadly the experience over many decades has been that too often scientists have asked many standard questions without much consideration of real users. The existence of non-adapted blanket fertilizer recommendations, after years of investment in trials of all sorts in country after country, is witness to such a process of inefficient field research (see examples in Chapters 2 to 4).

In contrast to the stylised models of plant nutrient cycles we see in soil ecology textbooks or the simplistic assumptions embedded in the design of many agronomic experiments, real farms are highly complex and diverse. For example, take one farm from Wolayta, southern Ethiopia (Figure 5.1). The picture shows how complex connections are, both within the farm, particularly the flows of resources (manures, household waste, ash etc) from the homestead to the garden plots close to the home, and beyond the farm, notably the flows to and from the market place. Such farmers' resource-flow maps (see examples in Chapters 2 to 4) are essentially highly complex systems models of a farm. Often they identify a wide range of flows and processes, with numerous connections and interactions, far beyond that which is normally modelled in standard computer simulations or budget models. By identifying key elements, they may be useful starting points for developing a simple model of the farm system, or designing experiments focused on issues identified as important.

With the challenge of sustainability on everyone's lips these days, the need to understand and exploit community- and farm-level dynamics and diversity and local users' perspectives, is even more acute than before. Finding ways that different types of farmers, with different views about soils and their

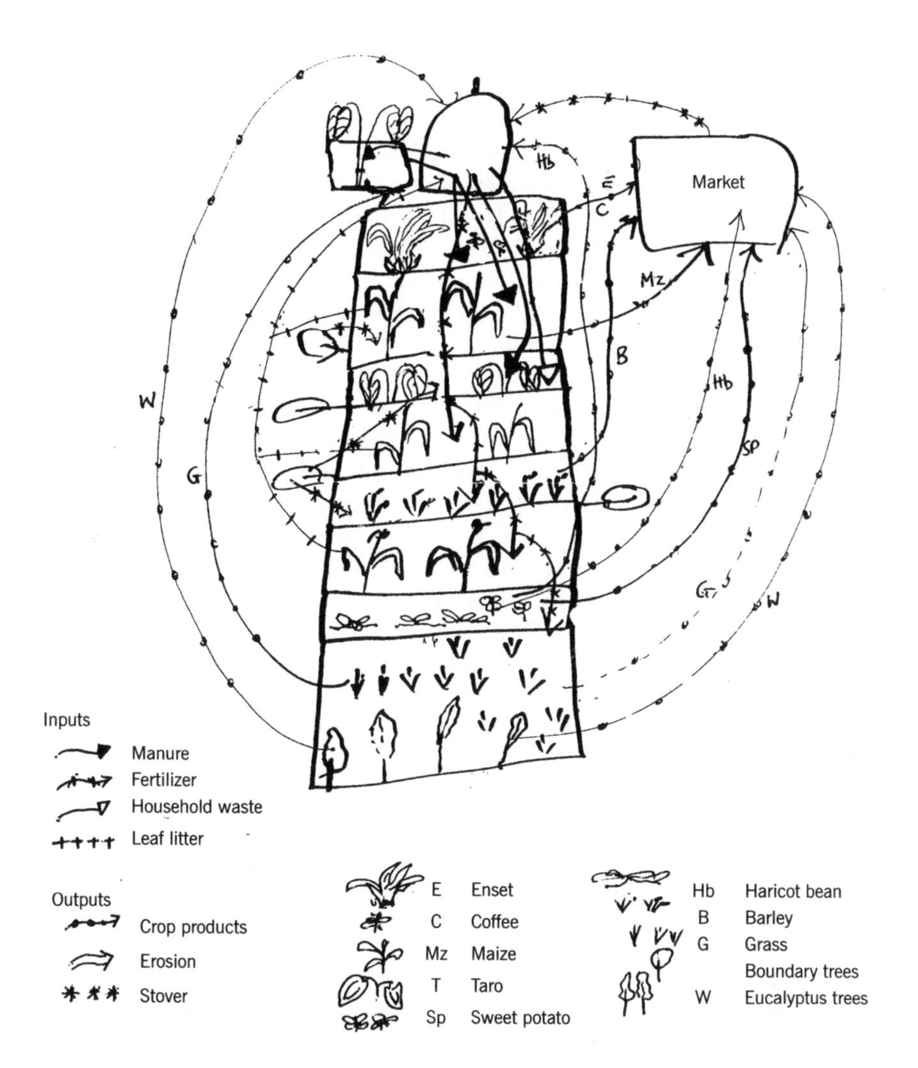

Source: Eyasu, 1997

Figure 5.1 *Resource flows in the farm of Hidotu Sankura, Wolayta, southern Ethiopia*

ecology and different opportunities to manage soil-fertility, can engage with conventional science and scientists is at the heart of the ISFM approach. The three country teams have developed and adapted various tools to analyse the prevailing diversity of soil-fertility and its management. For example, participatory tools for analysing landscape diversity and understanding the dynamics and diversity of farmers' soil-fertility management strategies have been developed. Equally, at the farm level, resource-flow mapping has proven to be an effective tool to visualize and analyse farmers' soil-fertility management strategies (Defoer et al, 1998).

FARMERS AND SCIENTISTS WORKING TOGETHER

The debate about farmer participation in agricultural research is now well established (see Farrington and Martin, 1988; Chambers et al, 1989; van Veldhuizen et al, 1997). Many researchers and development workers have also experienced the use of various types of participatory rural appraisal tools. However, in their eagerness to embrace the rhetoric of participation, the rationale for such partnerships is sometimes overlooked. Why is it that working together makes sense? For the field researcher there are many obvious reasons why a close relationship with farmers is essential. These include the need to understand the problem from users' perspectives; the need to frame questions and design experiments in ways that yield useful results; and the need to develop outputs with clear uptake pathways. The benefits of participation for farmers are, however, less obvious. Sometimes the designs and methods used may be useful in farmers' own research; sometimes new materials or technologies are offered for testing; and sometimes a genuinely new and useful output results. But all too often farmers are enlisted in scientists' projects, and the benefits from such engagements are not mutually shared.

However, experience from Mali, Ethiopia and Zimbabwe shows that, if sensitively designed, an interactive learning approach can provide benefits for both farmers and scientists, resulting in genuinely collaborative research and action. It appears that questions surrounding soil management are particularly well suited to this type of learning partnership. Why is this? Two factors can be considered: first, the significance of the issue and, second, the observability of the problem (Table 5.1). As has been shown in previous chapters, soils are clearly important to farmers, and fertility is often a key constraint in areas where farming has been continuous for some time. Also, soil-fertility is a complex and challenging problem that is often not immediately apparent, involving the interaction of soil, water, organic matter, macro- and micronutrients, with high levels of variability over time and space. Successful soil-fertility management demands a lot of practical knowledge of soil processes, many of which are invisible.

When soil-fertility problems are observable, farmers can establish cause–effect relationships. Indeed, farmers generally have a good understanding of the effect of soil-fertility on plant vitality and growth; a fertile soil allows the production of healthy plants that are resistant to pests and diseases and have a high yield. Farmers recognize, for example, that *Striga* infestation becomes significant under poor soil-fertility conditions, as a result of continuous cultivation without appropriate fertilizer application (Debrah et al, 1998). Although farmers have a lot of practical knowledge of soil-fertility and appropriate soil-fertility management, they generally lack understanding when it comes to less observable soil issues. This is the case for chemical aspects underlying soil-fertility processes. Aspects of mineralization of organic matter and the importance of appropriate C/N ratios in soils and soil amendments, and their effect on plant growth, are generally not well understood. The same holds true for plant-root infestation by nematodes, a pest that is less observable than *Striga*, but that is equally aggravated by inappropriate soil

Table 5.1 *Classification of problems according to significance and observability: some examples of soil issues*

Observability	Significance: Low	Significance: High
High	Surface crusting	Gully erosion *Striga* infestation Soil macro-fauna (termites, earthworms)
Low	Radionuclide concentrations Fossil deposits	Nutrient deficiencies Chemical and biological processes of soil-fertility transfer

Source: adapted from Bentley, 1994

management. One of the major reasons is that those processes take place below the soil surface, so they are not visible to the eye.

Farmers can often deal with those issues that are observable and important, perhaps with limited external support. In the context of soil issues, this category might include some aspects of soil-erosion control, such as gully reclamation (see Reij et al, 1996) or *Striga* infestation. But those issues that are important yet difficult to observe are perhaps the most challenging. Here, farmers and scientists can join together in fruitful partnership. Chemical and biological aspects of soil-fertility management, which may be combined with complex pest interactions, clearly fall into this category. In situations where everyone recognizes the importance of the issue but where observation is problematic, new tools to make things 'apparent' that are not visible with the naked eye are required.

ADAPTIVE EXPERIMENTATION AND JOINT LEARNING: EXPERIENCES FROM AFRICA

Participatory learning and action research approaches have been developed during the last five years in collaboration with a number of research partners across the three countries' research sites, including farmers' groups, NGOs and government research and extension organizations. At the local level a process starting with an initial diagnosis, and moving on to step-wise planning, experimentation and evaluation has taken place, involving inputs from individual farmers and scientists. Farmers' self-discovery through experimentation is central to the approach, with researchers facilitating the process. Learning is stimulated though feedback and sharing of experiences. From small beginnings, this increasingly yields wider impacts through the development of new networks and the creation of new organizational relationships to encourage ongoing learning and action. The basic approach, moving from diagnostic work to action research, is summarized below (see also Defoer et al, 1996; Mavedzenge et al, 1999; Figure 5.2).

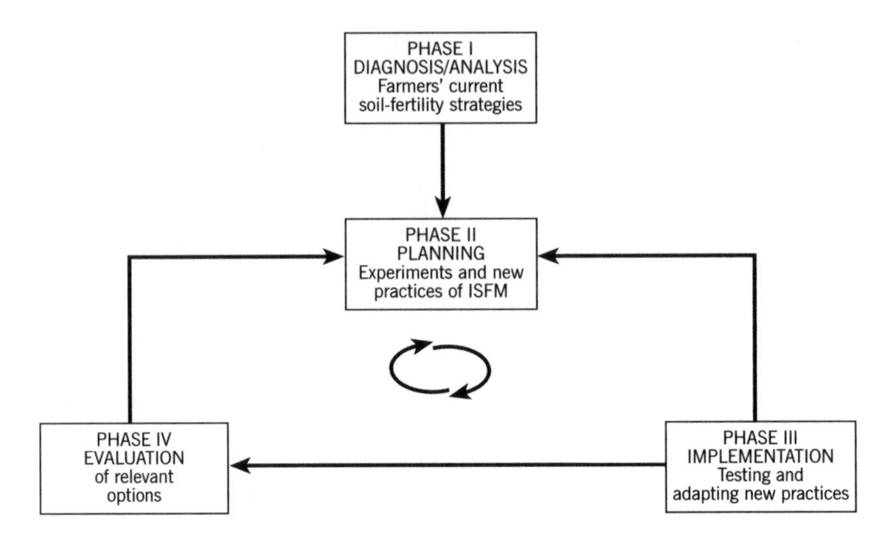

Figure 5.2 *From diagnosis to action research: phases in participatory learning and action research*

From diagnosis...

The process starts with an analysis of the land-use system at community level. First, an analysis is made of the existing landscape, the settlement pattern and its history and how natural resources are managed within the community's land-use system. Special attention is given to the diversity of soils and management practices. Second, farmers' information and communication networks are discussed and analysed. Understanding farmers' social relations and their types, sources and uses of information is important, since sharing information and experiences is key. Third, an analysis is carried out of the different soil-fertility management strategies prevalent within the community setting. The analysis focuses on differences between farmers in dealing with soil-fertility management and on the underlying factors influencing these differences, such as wealth, age, gender, access to resources, or external economic and political factors. This third type of analysis includes a classification of farms made on the basis of farmer criteria for appropriate soil-fertility management.

The joint analysis then continues at farm level, with selected farmers. They analyse the flows of resources on their farms, using resource-flow maps (RFMs) (see Lightfoot et al, 1992). These are useful to visualise and analyse their soil-fertility management strategies. An RFM presents flows of resources between major farm elements entering and leaving the farm. These flows are mediated by a variety of socio-economic factors. For example, market access or commodity prices affect some flows, while property rights and tenure affect access to and control over other resources. An analysis of such conditioning factors provides important insights for both farmers and researchers on the wider factors affecting the possibilities for farm-level action and sets the analysis in a broader context (see Chapter 6). Following this, areas for improvement are discussed, given the household's socio-economic conditions and overall strategy.

... to action research

Once different socio-economic, gender and age categories of farmers have analysed their soil-fertility management practices and resource flows, a number of problem areas are defined. Farmers in the area generally may feel these, or they may be more specific to particular farm classes or categories. The next challenge is to explore the array of potential solutions through searching for alternative intervention options. The search for new technical or management options often does not have to go far afield. Indeed, as people come together to discuss the results of the diagnostic studies and tours of farms are arranged, a range of different locally-generated ideas become apparent. In many cases these become the initial basis for testing and experimentation. Links with other sources of innovation can be made through visits to farmer innovators in other areas or to nearby research stations or experimental sites.

With a range of alternative options to test, farmers and researchers then design an experimental and monitoring programme together. Here again, the three country teams have put major emphasis on assisting farmers in making their own experimental designs and deciding what they want to be monitored (see Chapters 2–4). Table 5.2 lists a selection of the wide range of options explored by farmers across the sites.

Two parallel approaches to experimentation have emerged. Farmer research groups have often taken forward a more conventional experimental testing of options using simple designs and matrix scoring techniques for evaluation. For example, in Zimbabwe 20 groups have formed in the Chivi

Table 5.2 *Options tested and monitored by farmers and researchers*

Ethiopia	*Mali*	*Zimbabwe*
Crop-residue management on main outfields	Soil amendment experiments: compost, inorganic fertilizer, rock phosphate, cattle manure, cotton stalks and other residues	Soil amendment experiments for improving sandy soils: leaf litter, compost, ash, cattle manure, termitaria, inorganic fertilizer (and combinations)
Increasing manure and urine use efficiency, including reducing leaching/volatilization losses, bulking, improving application procedures	Improved fallow systems Improved quantity and quality of manure by bedding of animals and adding rock phosphate.	Improving sodic soils: river sand, termitaria, drainage lines, gypsum (and combinations) Composting pit systems and compost application rates
Low-labour-input composting techniques	Composting systems to increase recycling and nitrogen availability	
Inorganic fertilizer application procedures to increase use efficiency and reduce costs	Transportation of manure/ compost (carts, changing location of cattle pens, compost pits)	

Sources: FARM Africa, 1996; Defoer et al, 1996; IER, 1996; FSRU, 1996; Mavedzenge et al, 1999

district study area during the period from 1993 to 1997. Their average size is 7.7, with two-thirds being women, and they are usually based on existing groupings, often around extended kinship networks and residence clusters (Mudhara et al, 1996; Mavedzenge et al, 1999). Similarly in Mali and Ethiopia, individual farmers and groups have undertaken experimental fieldwork in collaboration with researchers in their fields. In all cases farmers have been key in designing and implementing the experiments. Evaluation of the experimental work was conducted by farmers, both individually and in groups, assisted by the teams.

In parallel, smaller groups of farmers have engaged in seasonal processes of planning and evaluating of improvements through resource-flow mapping. Initially, between 9 and 15 experimenting farmers across three categories of soil-management capacity (highly correlated with resource access) in each of the sites volunteered. Making use of the mapping technique, farmers plan their own activities. Although planning of activities is certainly not new for most farmers, it usually happens in their heads and in an implicit way. When planning takes place on paper something essential happens; sharing of ideas and initiatives, with members of the household, with colleagues or with external advisors or researchers becomes much easier. The written or diagrammatic information can also be used to evaluate farming performance, even after a long time, when the farmer may have forgotten exactly what has happened. Changes over years are assessed, and comparisons between different farm classes are made and judged according to farmers' own criteria for success. The resource-flow mapping technique can be considered as an equivalent to a farmer's diary. Visualizing the analysis of the farm and the planning of experiments and activities for the next season is an important result of the learning and action research process (Defoer and Budelman, 2000).

Alongside the farm-based experiments, activities of improved soil-fertility management, and processes of farm-level planning and monitoring, researchers have also collected data on soil changes, nutrient balances or yield levels (see Chapters 2 to 4). The process basically consists of capturing the information from the farmer-drawn RFMs, which includes data about the farm, the fields, livestock and household, and the resource flows. In addition, information is collected through more formal surveying and data from literature (see Defoer et al, 1998). RFMs assist effective data collection because omissions and mistakes are directly visible on the maps. Moreover, farmers do not only provide information, but actively participate in the analysis itself. Information obtained through quantitative analysis, based on data gained from RFMs, can improve the knowledge and perception of both the farmers and the researchers involved. A key stage in any joint research of this sort is for the different research actors to come together to discuss and interpret results. Feedback sessions are usually held during the harvest period when results are evident in people's fields, and later in the dry season when a more complete assessment is possible and when planning for the next season is undertaken.

Following analysis and discussion, the cycle of learning and action must repeat itself. In order to sustain the enthusiasm and energy that results in new insights and successful technological innovation and adaptation, new ideas and sources of innovation must be continually injected into the learning system. Thus, with new issues identified, the process of search, supply, intervention and monitoring must be repeated through an annual cycle. Across the three countries such a cycle has been continuous for a number of years now. There have been ups and downs, but, in some sites, there is growing evidence that, given limited external support, a sustained interactive approach to learning and experimentation can emerge.

NEW CHALLENGES

Once such a cycle of action research has been established, new challenges inevitably arise. During this process a number of questions have been faced. First, how can new approaches to interpretation and communication be devised which allow the languages and concepts of western and local sciences to be discussed? Second, how can issues of power be addressed in the context of new research and action relationships? How can new forms of professionalism be engendered among researchers and extensionists working in new ways with farmers? Third, how can a process of action, reflection and learning be sustained, and, indeed, scaled-up beyond case study communities and farms? How can new organizational forms be encouraged which continue to support these new relationships and networks?

Such questions are, of course, not new. They are central to the challenges posed by a more farmer-centred approach to on-farm research and extension (see Scoones and Thompson, 1994; Ashby et al, 1995; Budelman, 1996; Röling and Jiggins, 1997). How has the work in Ethiopia, Mali and Zimbabwe responded to such challenges? Not surprisingly, given the different settings and circumstances, diverse responses have resulted. But some general themes have emerged.

For example, issues of interpretation and communication are being explored through the development of visualization techniques based on a range of participatory rural appraisal tools. RFMs and ranking techniques, in particular, are easily taken up by farmers and provide some direct links to scientific analysis. In the case of Mali, RFMs drawn by farmers are not only important tools in participatory learning and action, but also allow the collection of information which can be successfully transferred into nutrient-flow analyses and balances. The outputs of such quantitative nutrient-flow analysis can, in turn, be useful in the process of participatory learning and action research. Hence, participatory action research and quantitative analysis are not necessarily contradictory, but can join together in a fruitful marriage (Defoer et al, 1998).[2]

Questions of power and professional attitudes are, of course, more difficult to tackle, yet are central to any transformation of the research and extension approach (Chambers, 1997; Blackburn and Holland, 1998). The participatory learning process implemented by the three country teams has been instrumental in bringing farmer innovations out of isolation. In Mali, for example, the improved validity of results obtained has helped convince more conventional researchers of the value of farmer experimentation in the process of technology development. Farmers now present the results of their experiments during regional research evaluation and planning meetings, increasing their stake in the process of technology development. A similar evolution has taken place in Zimbabwe, where, each year, during farmer-run feedback workshops, research and extension professionals are invited as participants by farmer research groups (Mavedzenge et al, 1999). Such 'reversals' (see Chambers, 1993) can have a powerful effect on outlooks and attitudes. However, participation is about how people interact, and personal change is especially needed for those with power and authority, such as researchers and development agents, working in rural societies. If such shifts are not reinforced by longer-term changes in formal training, incentive schemes and organizational change, then 'conventional professionalism' will inevitably be sustained. It should not be a surprise that this challenge remains an issue in all three countries.

Experience across all three countries shows that a participatory approach to soil-fertility management offers new insights and the basis for a potentially productive partnership between farmers and researchers. But these experiences have been on a small scale, involving relatively few farmers in detailed farm-level monitoring and a limited number of farmer research groups in plot-level experimentation. The case of Zimbabwe is encouraging; there has been a autonomous spread wherein farmer groups have spontaneously formed and demanded that they be involved in the research. But even these new groups rely, to some degree, on external support. For really effective scaling up, there remain many challenges of going beyond the small-scale project setting and institutionalising such approaches within government and extension services. Within the context of the decentralization of public services in Africa, farmer-based participatory approaches to innovation and technology development will become increasingly important.

Learning as a primary objective of the participatory approach has consequences in terms of time needed for a process to mature. Looking at the level of commitment required, one could argue that most research and development institutions in Africa cannot afford to become engaged in this way, and that the attention given to a small number villages is too much, and therefore too costly, especially given the limited, and often decreasing, capacity of government departments. But if spending time and effort in a few villages will eventually pay in terms of spin-off effects, such an approach may be justified. For example, farmers or village groups that have taken part in the initial stages of the process may assist neighbouring villages in experimenting and technology evaluation. In Mali, for example, the farmer groups that were actively involved in the learning and action research process now regularly organize field days for farmers of neighbouring villages. They demonstrate new practices and use their RFMs to explain changes in soil-fertility management. Farmers who have participated in such field days are increasingly setting up similar experiments (Defoer et al, 1999b). Spin-off effects, however, do not always come automatically. Specific attention should be given to stimulate the development of farmer networks beyond village boundaries.[3]

Implementing a participatory learning and action research approach has important implications for research and development organizations in terms of time and effort, skill development, networking with other research and development organizations, and willingness to change methods and approaches towards new partnerships with farmers. For this reason, there is a need to analyse critically the conditions required for organizations to support learning and action research processes. When such processes become part of an extension programme, farmers who have been actively involved for a number of years in the pilot villages may assist in setting up similar processes in neighbouring villages. At a higher level, facilitators of the initial process can train extension personnel so that they can conduct the process themselves and eventually provide support. Stimulating interactions between villages will therefore become a prime responsibility of the extension service. At a regional or national scale it is wise to select sites according to an understanding of agro-ecological zoning. But perhaps the greatest opportunity lies in the networks and organizations of farmers themselves (see Ashby and Sperling, 1994), which, given the right support in terms of training and organizational strengthening may, in the longer term, form the basis for a demand-led and participatory approach to soil management, and draw in scientists and other professionals on their own terms.

Chapter 6

Ways Forward? Technical Choices, Intervention Strategies and Policy Options

Camilla Toulmin and Ian Scoones

Introduction

This book has emphasized the importance of taking local contexts seriously. The case studies have revealed the importance of dynamics and diversity in all farming settings across all three countries. Some of the key findings are summarized below.

- Farmers have criteria for classifying soils, such as their workability, inherent fertility, suitability for certain crops, responsiveness to particular inputs and water-holding capacity. Farmers actively manage their soils in ways that build on these characteristics with the aim, over time, of thus improving their value for crop production.
- Africa's farming systems are highly diverse. This diversity is an important feature at all scales. At household level, farmers manage this diversity by different land-use practices, choice of crop and input levels. Diversity at the level of the village landscape is exploited through use, for example, of low-lying areas (variously known in the study areas as *bas fonds*, *vleis*, *dambos*), for moisture-loving crops, upland sands for millets and groundnuts, and gravelly slopes for grazing and woodland. Farmers value such diversity since it provides greater protection against the risk of crop failure. Diversity at national level is seen in the differential development of areas considered of high and low potential, infrastructural investment and ease of access to important markets.

- Farming systems are dynamic, responding to a range of internal and external pressures. Reliance on data from nutrient balances provides only a snapshot at one point in time, from which the direction and evolution of the farming system in question are hard to determine. There is no single pathway being followed by all farmers in a given site, but rather an array of directions being taken depending on their circumstances. Farmers are constantly seeking new strategies as problems arise or new opportunities develop.
- Farmers' management of soil nutrients depends on a range of socio-economic factors. Access to livestock, labour, credit and markets are of particular importance in explaining which farmers are best able to maintain and improve the fertility of their soils. The household remains central to the management of the farm and the mobilization of resources, such as labour and capital. However, other social institutions are also of great value in enabling farmers to negotiate access to obtain resources such as draft power, transport and credit. Such institutions also help farmers protect themselves from risk by the development of social networks through which help can be sought in times of need.
- Data on nutrient balances demonstrate a mixed pattern of accumulation and depletion, depending on plot, farmer and location. Overall, cash-crop land receives the major share of both mineral and organic fertilizers applied. Land sown to lower-value crops, such as coarse grains, shows a consistently negative nutrient balance. However, due to rotation of crops, cereal yields can often benefit from residual fertility stemming from the previous year. Location of plots is also important, with land close to the settlement or cattle pen receiving most nutrient inputs.
- Farmers in all sites have been affected by recent policy changes, such as structural adjustment, devaluation and land tenure reforms, as well as exogenous events such as drought. Such changes have brought about major shifts in returns to different crops, as well as the liberalization of crop marketing. Farmers are clearly responsive to such positive changes, which implies that governments have a range of measures by which to influence farmers' decision-making. This range of policy options is discussed in more detail below.

Given this diversity of local contexts and the complex dynamics of soil-fertility change, what is the most appropriate way to identify options by which to support more sustainable soil management by smallholders in Africa's more marginal farming areas? Such options must combine different elements: technical choices, strategies for intervention and a range of policy measures. The remainder of the chapter sets out to explore this question. First, the particular incentives – operating at a range of scales from the household to the village to the national context – for farmers to invest in soil-fertility management are discussed. These factors condition the range of technical choices that farmers make. The following section looks at the variety of ways farmers can and do manage their soil-fertility and the implications this has for intervention strategies by governments, donors and others. The success of such strategies are seen to be affected by agro-ecological conditions, household and community-level institutional arrangements, and national policies in a range of areas. It is these policy themes which are then picked up in the following section, when issues of devaluation and structural adjustment, credit, rural infrastructure, research and extension services, and land and tenure reform and decentralization are discussed. A wide range of policies, therefore, are seen to affect soil-fertility management. But how should initiatives directed towards soil-fertility issues be directed? The final section looks at the options, and suggests a strategy which recognizes the multiple dimensions of soil-fertility management, and a process which establishes a more open learning, adaptive process for the design, monitoring and evaluation of policy measures.

Diversity of agro-ecological setting and farmer practice

The conditions faced by farmers in sub-Saharan Africa are remarkably diverse. The research sites were chosen to investigate the significance of such diversity for farmer practice. Thus, the sites selected spanned a long transect in Mali from irrigated Tissana and millet-based Siguiné to M'Péresso in the cotton belt, from highland to lowland farming villages in Wolayta, southern Ethiopia, and from high potential Mangwende to Chivi in the low potential region of Zimbabwe. This diversity across the continent and within the different countries studied is mirrored at village level. The options available to poor farmers are much more constrained than those available to richer farmers who have easier access to labour, land, livestock, credit and cash. This theme of diversity can be taken further to farm level, where the management of soils within the farm varies very considerably between parts of the farm. Typically, certain fields tend to receive far greater concentrations of labour and nutrient inputs, while others are more extensively managed. Thus, in Zimbabwe, homefields close to the settlement receive most attention while outfields have limited applications; in Ethiopia, the *darkoa* plots supporting dense stands of enset receive regular supplies of manure and household waste, while further afield maize crops on the *shoka* plot have to make do with little or no amendments to the soil. Similarly, in land-extensive dryland sites in Mali, the *soforo* infields around the village are often black with dung by the end of the dry season and offer well-fertilized conditions for crops of maize and millet. By contrast, the large, shifting outfields, or *kongoforo*, produce a harvest for four or five years before being abandoned to fallow and occasional grazing. Where cotton has become important, the availability of chemical fertilizer modifies this pattern and enables more permanent cultivation of larger bush fields.

This diversity at different scales has very important implications for how best to support improved farmer management of soil-fertility. There are no simple messages regarding appropriate ways to manage Africa's soils. While depletion or 'mining' of nutrients may constitute a problem for certain fields, there are others which are accumulating nutrients. The research has helped identify the locations in which soil-fertility management has become a key issue.

Villages such as Siguiné in Mali can still manage a system of fallowing and nutrient transfers from grazing in crop land and thereby maintain grain yields, albeit at low levels. For them, low and erratic rainfall and uncertain prices are at least as much constraints on assuring food security as the fertility of their soils. The sites in Ethiopia demonstrate a relatively efficient system of managing and recycling of biomass and nutrients, although these are tightly constrained by limited access to markets, poor veterinary services and the high cost of mineral fertilizer. In Zimbabwe, the two locations demonstrated marked differences in terms of their productivity and market access. Mangwende is in a relatively high-potential area, with reasonable rainfall although poor soils, and is well connected to markets in Harare and Chitungwiza. By contrast, Chivi is in a semi-arid zone where frequent droughts reduce productive potential. Although there are good road connections to Masvingo to the north and South Africa to the south, there are fewer market opportunities.

The research on differences in soil-fertility management between socio-economic groups has helped identify households for which soil-nutrient decline constitutes a serious problem. Poorer households in all sites faced particular difficulties in maintaining the productivity of their farmland, due to very limited supplies of livestock dung and poor access to credit and inorganic fertilizer. Such difficulties are a reflection of their limited assets and capacity to mobilize resources more generally. Poor farmers in Dilaba, Mali and Chivi, Zimbabwe were additionally constrained by the disappearance of fallow land and lack of access to credit. This diversity has major implications for design of technical options, extension approaches and policy frameworks for improving soil-fertility management.

FARMERS' CONCERN FOR SOIL-FERTILITY MANAGEMENT: THE BROADER CONTEXT

As the case studies have shown, the capacity of different farmers to invest in improving soil-fertility management depends on a range of factors which operate at different levels. For the household, these factors include access to labour, livestock, land, capital and cash. In all sites, better-off farmers were much better able to invest in larger-scale use of organic and inorganic sources of nutrients. For example, in some sites, having some means of transport, such as a cart, is also critically important in allowing large quantities of biomass to be moved to and from homestead, *kraal* and crop lands. But there was also great diversity in how such inputs are used, since farmers do not spread them universally across all fields and crops. Instead, nutrients tend to be focused on smaller plots of land where higher-value crops (such as vegetables) are grown. In addition, rotation of land spreads fertility inputs over a series of years. For cotton farmers in southern Mali, nutrient inputs were positive for cotton in the year in which cotton is cultivated, but if account is taken of the subsequent two years of rotation with maize, millet, and sorghum, then over three years there is a net outflow of nutrients. Nutrient losses are much more significant on fields where lesser-value crops are grown and which are not a high priority for farmers. However, even here, farmers often make very effective use of the very limited quantities of nutrients available through careful placing and timing of such inputs in relation to rainfall and crop development. In all sites, farmers were experimenting with very small additions of mineral fertilizer at sowing or weeding time to gain some benefit at limited cost. At the village level, issues include the range of soils available, the overall pressure on farm land and its availability, access to grazing and forage resources, and the importance of labour flows between households, as well as location in relation to markets. At the national level, factors of relevance relate to macro-economic policy, input–output prices, access to credit, institutions and legislation regarding land tenure and its management, approaches to research and extension policy, marketing, and infrastructural investment.

Our research approach used a range of methods to discuss the importance of such factors at farm-household, village and national levels. Interviews with farmers regarding family histories and changes in soil-fertility management practices have brought out the combined importance of changes at household level (such as access to labour, livestock assets and off-farm income sources), and broader economic and institutional factors (such as major shifts in prices following devaluation, and land tenure reorganization). This interplay of household, village and macro-level factors helps explain the differences in performance between different farmers at the time of the research. However, it was also clear from the family and village histories that farmers' strategies are dynamic and continually adapting to a range of new problems, as well as the opening up of better opportunities both in agriculture and elsewhere. In some cases, major events such as drought in Zimbabwe have caused significant changes to the farming system due to reduced supplies of manure following heavy cattle losses. In others, a coincidence of good fortune at family level and the development of profitable options for crop production can enable particular farmers to improve their situation greatly. Thus, for example, the growth of horticulture in the Mangwende site in Zimbabwe has been the result of the growth of urban demand for vegetables and fruit in Harare. Those who have access to wetland areas, labour and reliable transport are able to cash in on this new market. Equally, some farmers in the irrigated rice village of Tissana, Mali were able to gain a large allocation of land within the scheme, because they had a large number of family members at the moment of land allocation. Such land has become increasingly valuable with the growth in vegetable gardening and marketing.

At the same time, the research also demonstrates that maintaining and improving the fertility of their soils is only one among a range of objectives which farmers are pursuing. Farmers are faced with many important choices relating to their farm enterprise, other economic activities and domestic commitments. The decision by farmers to invest effort and capital in improving the soils and productivity of their farmland will depend, in part, on pressures to do so, the perception that changes are necessary and the lack of other options. But it will also depend on their confidence that the returns will be worth it. Soil degradation and nutrient losses are unlikely to prompt changes in farmer behaviour until and unless they perceive clear benefits from doing so.

Exploiting soil capital is a rational strategy for farmers to pursue where the nutrients remain to be tapped and where clearing new land for such purposes is cheaper and easier than investing in purchase of inorganic inputs, or in the considerable amount of labour needed to recycle biomass through manuring and composting systems. Such extensive strategies are clearly apparent in Siguiné, Mali. Farmers are likely to moderate such 'soil mining' where they see declining yields and where they depend on this land for future harvests. Cotton farmers in M'Péresso, Mali are starting to engage in much more intensive nutrient management, following similar patterns as those in the more nutrient scarce sites of Ethiopia and Zimbabwe studied here.

Efforts to 'recapitalize Africa's soils' and to replenish soil-fertility need to take into account the broad-based nature of the farm household enterprise. This means that farmers have to weigh up the advantages of investing further effort in improved soil management in comparison with the much broader range of decisions they face, such as the gains from allocating labour to migration, more time spent on trading activities and the need to allocate resources for a forthcoming wedding. This finding points to the importance of setting soil-fertility management within a broader livelihoods approach (Carney, 1998; Scoones, 1998), a theme picked up later in this chapter.

TECHNOLOGICAL CHOICES

If the ultimate goal of intervention in soil-fertility management is to increase useful outputs for improving livelihoods, there are a variety of routes to do so, each with differing technological choices. Table 6.1 identifies four clusters of technological options. One focuses directly on the soil resource, by attempting to improve the capital stock, and thereby the service flows derived from it. The other three focus more on the flows themselves, either reducing losses, increasing inputs or improving efficiencies of nutrient use.

An approach aimed at increasing inputs and reducing outputs suggests a set of interventions, ranging from fertilizer application or soil conservation measures, through a range of biological management approaches, involving agroforestry, green manure production, legume use and so on. The details of such technologies are well known, and many combinations have been tested in field settings across Africa.[1] To different degrees, all options are evident in the different case study sites (see Chapters 2 to 4). Depending on the agro-ecology of the site, the asset base of the farmer and the institutional and policy context, different options have been picked up and used, singly or in combination.

A number of common elements can be found in the current research and extension activities being pursued in each of the countries studied. In most sites, extension recommendations are dominated by inorganic fertilizer recommendations. Where organic amendments are part of the extension package, they often recommend unfeasibly high levels of application rates.[2] In some sites, particularly as a result of NGO and research projects, a wider range of technologies for increasing nutrient inputs have been tested. For example, all sites have had composting pits tried out, alongside a range of agroforestry

Table 6.1 *Technology choices for managing stocks and flows of nutrients*

Managing stocks	*Managing external flows*		*Managing internal flows*
Increasing nutrient stocks	*Increasing nutrient inputs*	*Decreasing nutrient outputs*	*Increasing nutrient use efficiency*
P – Recapitalization through rock phosphate or other P additions N – Organic matter build-up	Inorganic fertilizer Manure/urine N-fixation (legumes) Fallowing, green manure, agroforestry etc Biomass import (leaf litter etc)	Erosion control Reduce leaching (mulching, deep capture through agroforestry, tillage systems) Reduce volatilization (manure/urine management)	Composting; residue recycling Water management Crop choice and management Fertility input combinations Input placement and synchronization

Source: Scoones and Toulmin, 1999

options. However, in most cases a simple transfer of technical design to a range of farmer settings has resulted in limited uptake. Thus, for example, in the lowland Ethiopian site, the multiple pit composting system recommended by the Ministry of Agriculture was quickly abandoned in favour of a simpler technology. Similarly, the alley farming techniques which had been promoted were also rejected by farmers, in favour of planting trees in homesteads and along field boundaries. In southern Mali, by contrast, considerable effort has been made by researchers and extension workers over recent years to help farmers develop organic fertilizer production through a variety of means, including composting, which provides organic materials for farmers with few or no livestock. In this way, research has aimed to address the obvious differences in assets and constraints faced by poorer and richer households.

Direct investment in the capital stock of a soil through recapitalization is an approach which has recently been widely advocated (Sanchez et al, 1997). In terms of technologies, this may involve the application of rock phosphate to enhance phosphorous stocks and the use of nitrogen-enriched fallows for nitrogen recapitalization. This has seen little application in the case study sites, although rock phosphate options are being explored in both Mali and Zimbabwe. However, the degree to which such additions may act to recapitalize natural capital or act as a cheap supplementary amendment (perhaps in combination with organic sources) is unclear. For most farmers, such technical solutions are not part of their current repertoire. Their means of increasing capital stocks has rather been through the painstaking, labour-intensive process of continuous fertilization and cultivation. This, as the case studies have shown, takes place largely in gardens and homefield sites where the large bulk of organic matter is placed. For other plots, the options for recapitalization are limited due to limited resources, so that farmers can only hope, at best, to maintain crop yield at a fairly low level.

All sites have experienced various attempts at soil erosion control. In many cases a top–down approach to technical design and implementation has resulted in a great deal of resentment. While farmers readily recognize the importance of stemming nutrient losses through erosion control, the imposition of fixed contour bunds or terraces has not been widely welcomed. Thus, for example, in Zimbabwe, the colonial attempts at soil erosion control became a focus for nationalist opposition during the liberation war of the 1970s, and in Ethiopia, after the fall of the Derg, many farmers dug up their terraces and replaced them with more flexibly-designed and less land-consuming alternatives (see Chapters 2 and 4). In the cotton zone in Mali, the CMDT in collaboration with researchers have incorporated soil-erosion control as a central part of their support to cotton farmers in Mali Sud (Hijkoop et al, 1991). This has had considerable success in reducing erosion losses and protecting the income streams of cotton farmers.

More sensitive approaches to land husbandry and soil conservation have emerged in the last decade, which draw more explicitly on farmers' own techniques for managing soils. These often combine the management of soil, water and nutrients in combination, rather than isolating soil erosion as the main issue (see examples in Reij et al, 1996). Combining physical with biological conservation measures and changes in tillage practice perhaps offers the most promising route for tailored interventions in this area. For example, in Zimbabwe research on conservation tillage has been combined with more conventional approaches to soil conservation and newer approaches to agroforestry and biomass management on-farm to come up with a variety of technical options (see Chapter 4).

Improving the efficiency with which nutrients are used is less often considered when choosing among a range of technological options, but is nevertheless vitally important (Nordwijk, 1998). For instance, application efficiency is sensitive to changes in the mix of inputs applied, where they are placed in relation to the plant, and timing of application in relation to moisture availability and plant requirements at different stages of growth (Woomer and Swift, 1994). A decline or improvement in use efficiency may make big differences in useful output, without any change in inputs, outputs or nutrient stock levels, and so can help to offset the effect of a decline in available nutrients and depletion of stocks for a period. For example, work in Zimbabwe has shown that a very limited amount of fertilizer input, perhaps combined with small amounts of manure, if placed in a particular way and at a time which maximizes uptake efficiency, can produce very significant yield responses, possibly far higher than a general application of an input in a blanket manner (Piha, 1993).

This suite of technological options for improving stocks and managing the flows of nutrients can be applied in a variety of ways. Four different intervention strategies can be pursued, either singly or in combination (Scoones and Toulmin, 1999).

1 One-time soil recapitalization strategy, especially of phosphorous using rock phosphate.

Table 6.2 *Key conditions for four intervention strategies*

Key conditions	*Recapitalization of nutrient stocks (one-time)*	*High external inputs (especially inorganic fertilizers)*	*Low external inputs (especially organic matter management)*	*Mixed strategy: integrated nutrient management*
Soil type and limiting nutrients	Appropriate for P limited soils, less so when N limited.	Any: fertilizer mix can be adapted to particular soil and plant needs	Most effective for N limited soils, through build up of organic matter	Any
Biomass availability	na	na	High	Variable
Labour requirements	High initial input	Low	High	Variable
Financial costs – credit or subsidy requirements	High	High	Low	Variable
Farmers' skill levels required – requirements for extension support	Low	Low	Medium	High
Need for good infrastructural/ market links for input supply	High	High	Low	Medium
Potentially negative environmental or health impacts	Yes	Yes	No	Limited
Intervention focus	Input importation; transport; implementation logistics; farmer and area targeting	Soil testing; fertilizer manufacture, blending; demonstration plots; input prices and markets	Appropriate technology development; training	Local experimentation; researcher–farmer participation; skills training; 'field schools'

Source: Scoones and Toulmin, 1999

2 High external-input strategy, based on the use of inorganic fertilizer packages.
3 Low external-input strategy, based principally on the use of locally available organic resources.
4 An integrated soil-fertility management strategy, making use of a range of high and low external-input technologies in combination.

A variety of conditions make any one or combination of such strategies more likely in any particular setting (Table 6.2). These include agro-ecological factors (such as soil type or biomass availability/productivity), household level conditions (such as labour or cash availability), and broader policy conditions (relating to prices, markets, credit, infrastructure and extension support).

The diversity of soils, cropping patterns and management practices within a single farm holding and the dynamic nature of soil-fertility and farming-system change sets a challenge for researchers and policy makers. Diversity of strategy at farmer level implies an approach to intervention which is highly localized and involves farmers in identifying, monitoring and evaluating different interventions for improving systems for soil-fertility management. This requires a flexible and responsive policy environment to support processes of technological innovation and investment in soil-fertility management. As the above table has indicated, a number of policy areas influence the likelihood of adoption of different options for soil-fertility management. It is to these issues which we now turn.

IMPACTS OF POLICY ON FARMER PRACTICE

Farmers in all three countries studied have been affected by a range of policy measures which have influenced crop choice, input use and broader livelihood strategies. Table 6.3 summarizes the range of policy interventions and their impact in the study areas. Five broad policy areas were identified in the case study research as having particularly significant impacts on soil-fertility management: structural adjustment (relating particularly to input–output pricing and marketing); credit; rural infrastructure; research and extension services; and land and tenure reform and decentralization. These are discussed in more detail below.

Devaluation and structural adjustment

In each of the three countries, there have been considerable changes in macroeconomic conditions and broader policies, linked most particularly to programmes of structural adjustment which have shifted substantially the terms of trade and incentives faced by farmers. However, the timing and nature of changes wrought by such measures have differed considerably between countries, as have some of the impacts. For example, Mali was one of the first countries in Africa to engage in structural adjustment from 1982 onwards, whereas Ethiopia has only recently begun to cut back on fertilizer subsidies. Overall, structural adjustment measures have led to the rising cost and reduced availability of chemical fertilizer, cutbacks in extension services and changes to crop marketing structures. However, while some farmers (such as those engaged in cash cropping, and those who depend on extension advice) have felt the impacts of such changes fairly immediately, others have been less closely affected because of their greater autonomy.

In Zimbabwe, the Economic Structural Adjustment Programme (ESAP) was introduced in 1991, resulting in changes to marketing, provision of services and input–output prices. These have brought about a major shift in use of inputs and choice of crops, leading to diversification and modifications to land-use management (Government of Zimbabwe, 1998). The impact of ESAP was particularly harsh since its implementation coincided with the

Table 6.3 *Impacts of policy on soils management in Ethiopia, Mali and Zimbabwe*

	Ethiopia		*Mali*		*Zimbabwe*	
	Policy context	*Policy impacts*	*Policy context*	*Policy impacts*	*Policy context*	*Policy impacts*
Input-output pricing	Liberalization and devaluation from 1991; subsidies on fertilizers removed 1997. Currency devaluation from early 1990s.	Real prices of both inputs and outputs increased, although profitability of inputs use constrained.	Structural adjustment from 1982 onwards, removal of subsidies for fertilizer and other inputs, abolition of credit programmes. Restructuring of CMDT, Office du Niger. Devaluation of CFA, 1994.	Liberalization of cereal markets, input prices up, food price increases, loss of access to credit.	Liberalization from 1991, with removal of all subsidies. Regular devaluations of Zimbabwe dollar since 1991.	Prices of inputs and outputs have risen, with price ratios remaining approximately similar. High value crops (eg vegetables, some cash crops) have become more profitable.
Marketing	Well-established informal markets; fertilizers sold through regional trading companies and private suppliers from mid-1990s.	Limited growth of private sector marketing, with negative impacts on prices for farmers. Reliance on extension or project supply of inputs through credit programmes.	Abolition of state cereal board, liberalization of cereal markets, including rice. Private input supply outside cotton growing area.	Growth in private trade, cereals and vegetables increased opportunities for farmers. Cotton remains a government monopoly through CMDT.	Effective privatization of marketing boards, and encouragement of private trade.	Growth of private trading has had positive impacts in the higher potential areas, but less so in lower potential sites where profit margins are lower. Here, growth of contract arrangements has occurred.
Credit	Range of informal credit options; bank credit limited; agricultural input credit focused on package programme.	Formal credit largely inappropriate for poorer farmers; package programme results in increased debt burden for poorer farmers and in marginal areas.	Formal credit available only to cotton farmers through village associations. No more credit provided by Office du Niger.	Cotton farmers continue to gain credit for input purchase. Rice farmers rely on private sector. Cereal farmers access informal sources, off-farm cash and NGO credit funds.	From early 1980s, limited availability of formal credit.	Credit a major constraint for many farmers.

Rural infrastructure	Poor road infrastructure, but significant investments since 1991.	Improved transport and marketing opportunities for remoter areas.	Gradual extension and upgrading of main road network. Little improvement in feeder roads.	Farmers near main roads face much cheaper transport to markets, many others rely on carts, donkeys, walking.	Since independence in 1980, major investments in rural road building.	Transportation and marketing is now easier, although costs have increased due to fuel price increases caused by devaluation.
Research and extension services	Research service undergoing reform, although current focus on limited technical options; extension dominated by package programme.	Much research inappropriate for poorer farmers, with resulting limited impact. Focus of extension service on package programme also excludes many.	Government research on agriculture under reform. Funding cut and shift to greater client orientation. Main focus has been raising yields of cotton and rice, from technical standpoint. Low potential areas receive no systematic attention, though occasional NGO projects.	Considerable impacts from research and extension for rice and cotton. Increasing attention to diversity of farmer strategies and messages. Very limited impacts of research for dryland cereals.	Research and extension switched emphasis to the smallholder sector following independence. However, significant budget cuts and a continued narrow technical focus have limited the ability of government services to respond to farmer needs.	Research has had limited impact on farmer strategies. The impact of a reduction in extension provision is unclear. Emergence of project-funded research and extension activities has had important, but limited, effects in certain areas.
Land tenure	Land owned by the state; no changes in formal land tenure regime expected.	Variable perceptions of land tenure security, although from late 1980s growing evidence of land-based investments. Various forms of land exchange, tenancy, share cropping etc persist.	Land taken as national asset post-independence, with free access for all. Current land tenure reforms allocate responsibilities to rural councils, who may attribute powers of management to village-level structures. Putting land to good use the basis for acquiring rights.	Lack of clarity regarding powers to manage and control access, most marked for common property grazing, woodland, unfarmed areas. Fairly secure tenure farmland through customary systems. Relations unclear between communes rurales (CRs) and village-level structures.	All communal lands remain state property, with rights to distribution now held by rural district councils and local traditional authorities, and recently land councils.	De facto land tenure involves a high level of security in communal areas. Recent emergence of an (illegal) land market in some areas, as well as land invasions

Table 6.3 *continued*

	Ethiopia		*Mali*		*Zimbabwe*	
	Policy context	*Policy impacts*	*Policy context*	*Policy impacts*	*Policy context*	*Policy impacts*
Land reform and resettlement	In past forced resettlement and villagization carried out; currently some discussion of further land reorganization.	Uncertainty over future intentions results in insecurity.	Uncertainty regarding rights and roles of CRs and village-level structures.	Future tensions and confusion likely between powers of CRs and customary structures at village level.	Land reform has been a central pillar of post-independence policy, but until 2000–2001 limited progress had been made.	Land reform and resettlement has had limited impact on farmers in most communal areas.
Decentralization	Regionalization a central policy thrust.	Opportunities for tailored regional programmes and policies, although limited capacities and continued influence of national directives undermine options.	Process of preparation over last 5 years, elections held May 1999. Heavy donor pressure, political control and service delivery the main reasons for decentralization.	Too early to judge. Clarity needed regarding CPR management, rights and roles of CRs and village-level structures.	Decentralization to rural district councils has been ongoing since the 1980s. Lower level development committees at village and ward level were established.	Much of the process of decentralization has involved deconcent-ration, with resulting reductions in services offered at local level. Village and district authorities have relatively little authority and influence.

severe drought of 1991–1992, during which a high proportion of livestock died. Hence, for example, farmers in the drier area of Chivi lost 60 per cent of their cattle holdings, drastically cutting the availability of manure for maintenance of soil-fertility. As a result, Chivi farmers have greatly reduced capacity to maintain soil-fertility and must concentrate the limited amounts available on their most important crops. The liberalization of grain marketing means that farmers are no longer obliged to sell their maize at a fixed price to the Grain Marketing Board, which now just sets a floor price at which it can constitute essential stocks in case of future shortage (FSRU, 1998). As a result, farmers are exposed to potential gains, but also risks associated with the private trading sector. Private traders are now starting to engage in provision of inputs to farmers for particular crops, thereby establishing an alternative input and marketing channel. Contracts with traders are of greater importance in the higher-potential Mangwende for cotton and vegetables, but, even in Chivi, some farmers are now regularly growing peppers and other vegetables for purchase by traders. In general, less grain is now being grown, with farmers resorting to vegetable gardening which has higher returns per hectare (FSRU, 1998). Limited nutrient inputs are focused on gardens in homefields, or those being established along river banks. With fewer livestock available, there has been a return for many to hoe agriculture, and greater attention paid to incorporation of plant matter within ridge and mounding systems (see also Chapter 4).

Zimbabwean farmers' perceptions of the ESAP are mixed, as might be expected, depending on their access to resources and the opportunities they can gain from changes in prices and markets. In Chivi, farmers' perceptions are universally negative:

> *Those who are richer get richer, and those who are poor become poorer... ESAP is like AIDS – it kills you slowly, but surely... farmers cannot benefit fully from market liberalization because the markets fix prices at levels which are disappointingly low...* (FSRU, 1998, p458)

In Mangwende, there is more of a contrast in view, given the higher level of agricultural potential and marketing opportunities available:

> *ESAP is about encouraging people to work hard to build wealth... removal of price controls and competition amongst businesses can be beneficial if it leads to reduced input prices and better producer prices, as is happening with cotton... ESAP means a more difficult life for farmers because input prices are rising faster than producer prices...* (ibid)

In Ethiopia, the impacts of structural adjustment on fertilizer prices are only now starting to feed through to farmers, for whom fertilizer prices have risen significantly in real terms (Croppenstedt et al, 1998). For farmers in Wolayta, these changes have rendered even more inaccessible their access to purchased inputs which has been in steady regression since the closure of the WADU project in 1981 and the withdrawal of credit-based fertilizer support. In the last few years, the Ethiopian government has been strongly promoting and extending the fertilizer package developed with the SG-2000 programme. This may provide the possibility of accessing assured supplies of inorganic inputs at reasonable cost for better-off farmers with sufficient land and good market access. However, as discussed in Chapter 2, the SG-2000 package raises serious concerns when spread to poorer farmers and drier, risk-prone areas.

In Mali, structural reforms were begun in the mid-1980s and included liberalization of cereal prices and the abolition of fertilizer credit schemes. Many farmers in the dry cereal-producing areas had never had easy access to such inputs since they were not in an area considered a priority for rural development interventions. However, farmers in Dilaba did have the opportunity to tap into project-supported credit and chemical fertilizers during the 1980s. For rice and cotton farmers, structural adjustment and devaluation of the CFA franc have had a much greater impact because of their engagement in market activities.

Devaluation of the CFA franc in 1994 has had a largely positive effect on farmers in Mali. Prior to 1994, cotton farmers were facing stagnant prices and rising input prices. The CMDT had limited processing capacity and had established a quota system to keep a ceiling on cotton production. The devaluation coincided with rising world market prices for cotton, such that very substantial benefits could be gained from expanding area under this crop. With the CMDT investing in new ginning capacity, and a lifting of quotas, the area under cotton has grown from 200,000ha in 1993 to nearly 500,000ha in 1996 (Jeune Afrique, 1998). While input levels in the first year after the devaluation were somewhat depressed, subsequently farmers re-established former practices. Thus, for example, average net returns per hectare of cotton rose by a factor of 3 to 5 over the period 1993–1996 (Giraudy and Niang, 1996). Prices have also risen for cereals, particularly maize, given high levels of demand from the neighbouring Ivory Coast. Not only the larger, better-equipped farmers, but also the poorer small-farm households have been able to take advantage of these opportunities and expand the area cultivated, gaining a substantial increase in income per hectare because of the large hike in cotton prices paid to farmers. In the irrigated rice zone of the Office du Niger, devaluation has also brought about increased demand for domestically-produced rice, since imported Thai rice has become more expensive. As a result, farmers have seen a rise in returns averaging 10–35 per cent per hectare, despite an increase in input prices (Breman and Sissoko, 1998). Those farming households with livestock holdings have also done well following devaluation, given rising prices and increased demand from coastal West African states.

Structural adjustment has brought other changes to certain farming areas of Mali, particularly those covered by the CMDT and the Office du Niger (as described in more detail in Chapter 3). The Malian government has been required to privatize certain functions formerly carried out by these parastatal organizations, such as provision of veterinary services. The rapid withdrawal from veterinary service provision in 1995 led to heavy cattle losses from disease and damaging impacts on work oxen numbers in the CMDT zone. The CMDT has also been forced to become more transparent in terms of its pricing policy, and the share of cotton profits to be received by each of the three parties – farmers, the state and the CMDT. The programme to restructure the Office de Niger has brought changes in the responsibilities and rights of tenant farmers, providing improved security of tenure, and greater freedom over crop choice and where to market crops. Contract farming of certain vegetables (such as tomatoes for canning) is also now starting to develop in the irrigated lands of the Office du Niger, with traders providing access to credit, inputs, markets and technical advice.

Credit

Throughout sub-Saharan Africa, farmers have limited access to formal financial systems through which to gain access to credit and in which they can invest their savings. The establishment of such systems faces serious constraints which include low levels of income, poor infrastructure, high transaction costs and limited collateral with which to hedge against risks of default (de Groote, 1995). As a result, savings at village level are frequently held in the form of livestock, and farmers rely heavily on informal sources of credit from neighbours and family and earnings from migration and off-farm activities for purchase of equipment and farm inputs. In the research sites, systems to gain access to credit have been undergoing change as a result of structural adjustment and other measures.

Across the study sites, only in the CMDT area of southern Mali is such access more or less assured, though even here the CMDT has passed on responsibility for managing the loans to the National Agricultural Development Bank. In all other sites, while projects may occasionally set up and supply credit to a limited area, in most places farmers must rely increasingly on their own resources or on private sector loans, usually associated with cropping under contract to a commercial buyer. Thus, for example, for rice farmers in Mali, credit is no longer supplied by the Office du Niger, and farmers must rely on private sources with which to purchase inputs and equipment. In the southern Ethiopia research sites, most farmers rely on informal networks through which to gain access to loans. Bank credit is in very short supply and dominated by a few suppliers requiring significant collateral arrangements. Agricultural credit is otherwise limited to the extension programme which offers a fixed package of inputs on credit to farmers with strict systems for repayment. Poorer farmers are in particular difficulties as a result, having very limited networks for raising cash while being unable to make effective use of the credit package. As a result, rising levels of default and debt are affecting the poorer and more marginal farmers in the area. In Zimbabwe, formal credit was offered to all communal area farmers on reasonable terms in the 1980s through the Agricultural Finance Corporation. However, default levels, particularly in drought-prone areas such as Chivi, were very high and the facility has been withhdrawn. Today, getting access to credit from government or private bank sources is increasingly difficult, in part because of the stringent conditions applied. Informal lending and savings arrangements are therefore the primary way by which farmers raise cash.

Rural infrastructure

Easy access to markets can provide a major spur to farmers to intensify agricultural production and investment in land improvement (Tiffen et al, 1994). Such effects stem from the impact of a broader array of demands for goods and services, and reduced transport costs bringing better margins on input and output prices.

In southern Ethiopia, the challenges set by the mountainous landscape to road building, combined with low levels of investment, have meant that rural infrastructure is poorly developed. However, since 1991, major improvements to the main road through the study area has brought shorter journey times and more assured access throughout the year. This has been particularly important for the lowland study site where, in the past, travel times to the main market town of Soddo were considerable. Today a greater range of motorized transport opportunities exist at much lower costs. Nevertheless, much transport to market continues to be done by foot and donkey, with people often trekking for many hours to reach a market. The profusion of markets, occurring on a daily basis (and sometimes with day, evening and even night markets), is a characteristic of the Ethiopian rural economy. Such markets give traders the opportunity to make small margins by selling and buying in different places. This source of off-farm income is particularly significant for women. With the increasing access, larger market centres such as Soddo have become more important, sometimes displacing smaller, local markets. This has a number of consequences. For those purchasing and selling goods, improved infrastructure has reduced transport costs with the result that both consumers and producers benefit. However, the reduction in trading opportunities in the smaller markets may have impacts on women's livelihood options, as it is often men, with access to cash, who can afford the transport costs to the larger market centers, with the result that trading activities become concentrated among those with capital assets.

In Mali, maintenance and improvement of the principal roads from Bamako to Sikasso, Segou and the north-east of the country takes up the major share of the roads budget, with very little done to improve feeder roads away from the main arteries. In the CMDT zone there is assured transport for the collection of cotton and delivery of inputs. For Tissana and Siguiné, the large market town of Niono is relatively close and provides many opportunities to diversify crops and income sources. Even in areas much further from markets, people often walk very considerable distances to market. But these distances limit the kind of crop which can be grown for sale, since it must withstand the heat and jolting during the journey, as well as increasing the time and costs involved.

In Zimbabwe in the 20 years since independence there have been major investments in road building, including roads within the communal areas, making transport and marketing considerably easier. There has been a resultant growth in bus and lorry traffic in the communal areas, with competition between operators driving prices down, although this effect has been offset to some extent by higher fuel costs stemming from currency devaluation. There are some differences in the level of rural road infrastructure between the study areas, with Mangwende being particularly well served with good connections to Harare and Marondera. Farmers regularly hire transport in groups for the transport of their produce. For vegetables in particular, timely delivery is key in order to gain the highest prices. In Chivi, the road networks have significantly improved in recent years. However, the distance to major market towns is larger than in Mangwende, with Masvingo being the most important. The road to South Africa which runs through Chivi provides important opportunities for trading both agricultural produce and other items. The tourist traffic along the road also provides a ready market for a range of produce, including craft items.

Research and extension services

Research and extension service provision has undergone major changes over the last decade in all three countries. In part this has been due to structural adjustment reforms which have reduced the expenditure levels in government services. But, in addition, there have been some other changes in focus, with a gradual realization of the need to gear research and extension support to the needs of small-scale farmers. In all three countries this has been a slow process, but through interaction with a range of project-based initiatives, both within and outside government, some changes are being seen in the broader national agricultural research and extension strategy in all three countries.

Since the 1970s, agricultural research and extension in Ethiopia has been associated with a technical package approach. A range of programmes have advocated different technical solutions to small-scale farming problems, but with relatively limited impact. While there has been important research, particularly on new varieties, this has had a limited impact on farming livelihoods in areas such as Wolayta. One criticism is that research has been largely focused on cereal crops for the highland areas. This is of clear national importance, but it has resulted in other agricultural production systems – including particularly the enset/root crop and pastoral systems of the south – being largely neglected. However, over the last few years the national research services have been undergoing signifcant reforms. With the formation of the new Ethiopian Agricultural Research Organization, a new and broader mandate for agricultural research has begun to be defined.

In terms of extension outreach, however, the focus remains on a package approach. This has been reinforced by the national adoption of the improved crop-fertilizer-credit package originally promoted by the NGO SG-2000. While there are undoubted benefits of the package under the right conditions, this does not suit everywhere and everyone. In Wolayta the previous experiences of the WADU programme during the 1970s provide an important lesson. When subsidized credit was available and the infrastructural, marketing and other requirements were covered by the project, significant gains were achieved in cereal production. However, this was at the expense of the traditional root-crop system, and when the project withdrew in the early 1980s, cereal yields collapsed and farmers had to return to a focus on their previous mixed strategy of combining enset with root crops and cereals. Not suprisingly, the impact of the government's agricultural extension strategy in the Ethiopia study areas has been mixed, with richer farmers in the higher rainfall areas being able to make good use of the package. However, poorer farmers by and large have not benefited and many, following poor rainfall years, have gone into debt (Carswell et al, 2000). The limitations of the approach, however, are now beginning to be recognized and a strategy for addressing the problems of poorer farmers in more marginal areas is being developed, with an extension of the range of extension packages to such issues as soil and water conservation and livestock management.The agricultural research sector in Mali largely consists of the Institut d'Economie Rurale (IER) which, despite its name, is mainly technical in focus. Agricultural research has paid greatest attention to the problems faced by cotton farmers in the CMDT area and irrigated rice producers in the Office du Niger. Recent structural reforms to the IER have led to cuts in funding and greater reliance on a range of clients, particularly donor agencies requiring short-term research support for their field projects. At the same time, IER has moved towards a regional model, with greater responsibilities allocated to the 5 regional research centres (of which Niono and Sikasso are part). At the regional level, consultation committees have been set up in the hope of establishing closer communication between researchers and farmer groups in that area. However, it is unclear how effective these have been in influencing the direction and methods followed by researchers. Rice and cotton farmers have thus received the lion's share of research expenditure. In both cases, this seems to have brought considerable improvements in yield when combined with favourable market conditions. Much greater attention is now being paid to differences between farm households in the CMDT region, so that the difficulties faced by small, poor households can also be addressed.

Rainfed cereal areas, by contrast, have not been the object of concerted research and extension work, with provision of technical advice and credit the result of limited-duration programmes such as the Programme National de Vulgarization Agricole, or the presence of an NGO field project. Despite this neglect, farmers outside the main cash-crop zones have made very considerable changes to their cropping systems – taking up new varieties, adopting ox-drawn ploughs, and experimenting with inorganic fertilizer – based on their own capacity to learn, experiment and share lessons. While earlier research focused on the development of technical packages for farmers based on inorganic fertilizer, much greater effort has been made in the CMDT area over the last 10 years to promote more sustainable management of soils. This is being done by work on anti-erosion measures combined with better use of organic materials such as crop residues and manure. Private sources of extension advice and input supply are also growing, related to particular crops being cultivated under contract to a commercial buyer. Thus, for example, in the Office du Niger, irrigated tomatoes are being grown in the dry season for a canning factory, with seed, other inputs and advice being supplied to the farmers by the buyer.

In Zimbabwe, government policy towards agricultural research and extension switched from a concentration on the large-scale commercial farming sector to smallholders following independence in 1980. However, this transition has not been easy. The available technologies and the style of research which predominates in the Department of Research and Specialist Services (DRSS) of the Ministry of Agriculture has resulted in a relatively limited uptake of research results in the communal areas. This problem has been compounded by continuing budget cuts, particularly in the period following structural adjustment. This has limited the ability of station-based researchers to get to the field and interact with farmers. Nevertheless a number of project-based initiatives have in recent years started a more farmer-focused approach to agricultural research. The Farming Systems Research Unit (FSRU) in DRSS, for example, has developed a farmer-led participatory research approach in Chivi and Mangwende (see Chapter 5), with many experiments focused on soil-fertility issues. This has been complemented by more technical research both within DRSS and at the University of Zimbabwe supported by the Rockefeller Foundation-funded SoilFertNet. The result has been the exploration of a range of technologies and management practices which are more in line with farmer demands, and, in some instances, a quite rapid uptake of new practices.

In Zimbabwe, the links between research and extension have been relatively weak, except through more informal contacts at the local level. Agricultural extension, run by the Ministry of Agriculture department Agritex, has been hit by similar budget cuts as research. This has meant a decline in extension coverage over the 1990s, particularly as a result of reduced mobility due to mileage restrictions. In rethinking the role of extension in the country, a number of important lessons have been drawn from past experience. First, the top–down technology driven package approach which has dominated Zimbabwean extension support since the 1930s has been criticized as not really addressing farmer demands. Second, the high costs of broad field-level extension coverage (with the aim of having an extension worker in every ward) have been seen to be unfeasible given government budget limitations. In particular, the training-and-visit approach, originally recommended by the World Bank in the period following independence, has come in for a lot of criticism as being both too technically oriented and too costly. Drawing on a variety of experiments in alternative types of extension approach, particularly emanating from Masvingo Province (Hagmann et al, 1998), Agritex is now exploring more participatory approaches which redefine the role of the extension worker and the linkages between farmers, researchers and the extension system.

Land and tenure reform in the context of decentralization

Land tenure security is often seen as a critical factor in ensuring effective long-term investment in natural resources. A variety of land tenure reform initiatives have been started across the three country case studies. Various forms of land reform and resettlement have also been significant in affecting the distribution of land resources across the study areas. Land and tenure reform is occurring in the context of broader administrative changes associated with programmes of decentralization.

Farmers in Ethiopia have witnessed major upheavals due to land tenure changes and villagization, which have generated continued uncertainty over their land holdings. The land reform policy implemented by the Derg regime during the 1970s resulted in a major reallocation of land to farmers who were previously reliant only on tenancy arrangements with landlords under the feudal system. While this disrupted the strategies of richer landlords, it did provide new land for those previously unable to farm for themselves. However, despite the reforms significant inequalities in land holding persist, with former landlords often repurchasing or contracting land from poorer households. Today a huge range of land-holding arrangements exist, including outright ownership, share cropping, contracting and different forms of tenancy. While many of these are notionally illegal, the range of informal institutional arrangements surrounding land are an important aspect of the local situation.

Resettlement policies to less densely populated areas in the lowlands have been important in Ethiopia since the late 1960s. For example, the lowland case study site was established as a settlement scheme in 1971 as part of the WADU integrated rural development project. The allocation of 5ha plots to former tenant farmers provided new opportunities for agricultural livelihoods, although the consequences of moving to a lowland area with different agro-ecological conditions and high levels of both human and livestock disease incidence caused many problems. During the 1980s, the policy of villagization caused further changes in land-holding patterns. Villagization was aimed at rational planning for increased production, combining private and collective farming arrangements. The broader political effect was to increase the state's political and economic control over peasant farmers. Farmers in lowland areas of Wolayta were forced to move into villages and to leave behind the plots of higher-fertility *darkoa* land which they were in the process of enriching. Fortunately, they were to return to their home sites following the fall of the Derg in 1991 and reinvest in their garden areas. Resettlement and land reorganization have not been major features of land policy under the current government, with the land redistribution efforts in the Amhara region in the mid-1990s not being repeated elsewhere. Movement between sites instead tends to be more spontaneous, with families resettling to new areas (particularly in the lowlands) through informal connections with relatives and other contacts.

In Ethiopia there is continued debate at national level concerning further changes to land tenure (Dessalegn, 1994). The government insists that all rural land will remain the property of the state, but that this should not undermine security of land holdings and customary inheritance arrangements. However, given past government policies – particularly of land reform and villagization – many farmers remain sceptical about government intentions and a sense of uncertainty continues (Worku, 1998). Yet despite this, farmers continue to invest in long-term soil-fertility improvement in their garden sites, and the growth in on-farm tree planting in the area suggests that farmers are not thinking only for the short term (Carswell et al, 2000).

Regionalization, involving a form of political and administrative decentralization, has been a central plank of the post-1991 government's policy. While a process of devolving certain powers to the different regions which make up the country has occurred, it is unclear as yet how far this will lead to major changes in agricultural policy, given the continued political and administrative tensions between central government and ministries at the federal level and the new regional administrations, where capacity and political authority often remains weak. This is particularly apparent in the southern region which is made up of a wide range of disparate ethnic groupings (Young, 1996).

In Mali, the ongoing process of decentralization and land tenure reform provides an evolving landscape within which farmers must plan their strategies. Control over land currently being farmed seems relatively secure, with customary rules predominating (Lavigne Delville, 1999), but there is much greater uncertainty over management of collective resources such as grazing, woodlands and water (Joldersma et al, 1996; Hilhorst and Coulibaly, 1998). These common resources provide a variety of important products, including pasture for livestock which transport nutrients from bush to cropland. In addition, it is also not clear how the powers of the newly established rural communes will relate to customary structures at village level. While the tenure reforms (*Code domaniale et foncier*) propose allocating firmer rights to community groups to control access to resources within their territory, this may conflict with the claims that rural communes may wish to assert over resources now under their formal responsibility (IIED, 1999). Where communes see the allocation of rights as an important source of income and patronage, serious difficulties may arise over the prerogatives of village and commune-level structures. Given that the new *Communes Ruraux* were elected in May 1999 and are currently in the process of establishing themselves, it is too early to judge their likely impact. However, it should be remembered that the rationale for decentralization in Mali (and many other countries) has been based far more on service delivery and political considerations, rather than ensuring more effective management of land and natural resources.

In Zimbabwe, debate regarding land and resettlement has been underway since Independence in 1980. After an initial flurry of activity in the early 1980s there has been little action on the ground until the last few years (Moyo, 1998). From the late 1990s, the issue gained greater attention from both government and donors, with a new phase of the resettlement programme approved in 1999. During 2000, however, the politics of land reform changed dramatically, as the ruling party and the 'war veterans' began to focus their electoral campaigning on the land issue. A range of land invasions followed, and an attempt by the government to institute a 'fast-track' resettlement policy. The degree to which such resettlement programmes will reduce land pressure in the existing communal areas, however, remains uncertain. Nevertheless a number of studies show that, if the appropriate support is given to new settlers and resettlement takes place in areas where productive agriculture is viable, then the potentials for a resurgence of smallholder agriculture in Zimbabwe are great. In practice, though, the consequences to date for farmers in the Chivi and Mangwende study areas have been minimal. Relatively few have been able to sign up for resettlement programmes, which have been targeted largely at richer farmers with Master Farmer Certificates. Instead, more informal processes of resettlement are occuring with farmers moving, particularly from dryland Chivi, to other parts of the country (notably Gokwe and the Zambezi valley area) where land is more abundant, and, more recently, nearby commercial farms have been invaded and occupied.

Parallel debates centred more on land tenure reform and decentralization are also ongoing. Following growing concerns about land tenure and rural administration issues in the communal areas, a commission of enquiry was established which suggested a range of changes to land administration and tenure issues (Government of Zimbabwe, 1994). The commission recommended that village areas be granted tenure rights on a collective basis, combined with a review of the village and ward committee system which was widely regarded as inadequate. A greater role for 'traditional' leaders and customary arrangements was argued for. However, the implementation of the recommendations has been fragmentary, with particular rural political contexts largely affecting change at the local level. With the political hiatus of the last few years, little progress has been made on the ground. Decentralization to Rural District Councils has also been seen as a route to more effective agricultural and land management in the communal areas, but, again, the capacities of district councils to do much without significant budgets and in the context of joint responsibilities with central line ministries has been very evident.

POLICY OPTIONS AND STRATEGIES

A range of policy measures, therefore, is seen to affect farmer management of soil-fertility. However, rarely has soil-fertility management itself been the main target of such policies. Rather, impacts on soil management tend to be the cumulative results of a series of interventions whose focus and interest lie in other fields. As Scherr (1998, p2) notes, 'policy makers typically consider soil quality not as a policy objective in itself, but as an input into achieving other policy objectives'.

The launch of the Soil-Fertility Initiative (SFI) by the World Bank, FAO and other donors was intended to remedy this weakness (World Bank, 1996). One objective of the SFI has been to identify more clearly those policies which are likely to have an impact on soil-fertility management, where overlaps and contradictions currently exist. National soil-fertility action plans are intended to promote greater coherence in policy-making by bringing together those bodies with responsibilities which are likely to affect soils, and focusing on ways to encourage longer-term investment in improving soil quality. Thus, for example, the steering group for the elaboration of national plan of action to address soil-fertility management in Mali comprises representatives from the Ministries of Planning and Rural Development, as well as from the CMDT, the Office du Niger, the National Environmental Action Plan, the agricultural research sector, the World Bank and the Dutch Embassy.

It is important to examine both the content of policy (statements and instruments) as well as its process (elaboration, implementation and review) and to try and understand differences between what is said, and what actually happens (Keeley and Scoones, 1999; Mayers and Bass, 1999). It is clear that:

> *The policy-making process is by no means the rational activity that it is so often held up to be... policy research over recent years suggest that it is actually rather messy, with outcomes occurring as a result of complicated political, social and institutional processes.* (Juma and Clark, 1995, pp128–129)

Several points emerge from examining the policy process in the three countries studied. First, policy changes have stemmed largely from the demands of a range of international actors, rather than from internal debate and consultation among stakeholders at national and local levels. In many African countries, agenda-setting tends to be done within a very limited group of people, largely within government, with little or no public consultation, making it much more difficult to introduce alternative views and perspectives. The World Bank, IMF and major donors have also been able to maintain a highly influential role in policy directions because of the high level of indebtedness amongst many sub-Saharan African states, and their consequent dependence on development assistance. Governments remain heavily reliant on, and reactive to, new international initiatives (such as the Convention to Combat Desertification, and the SFI), often in the hopes of gaining renewed funding rather than necessarily feeling committed to the objectives of the initiative in question.

Second, the data which drive such policy debates are heavily reliant on a very limited number of documents, which are repeated time and time again. These include the FAO survey of 1990 (Stoorvogel and Smaling, 1990) and research which attempts to put an economic estimate on soil-nutrient losses (van der Pol, 1992; Bishop and Allen, 1989). Such data provide simplistic messages regarding the significance, diversity and dynamics of soil-nutrient management at farmer level, and also lead to a blinkered view of the means by which to encourage more sustainable agricultural practices generally. The end result is often a set of policy statements which bear only partial relation to what can be seen at farmer level, but which many are unwilling to contest.

Third, the current policy focus on soil-fertility management needs to be understood in the light of the structural adjustment measures undertaken by African countries under IMF and World Bank guidance in the 1980s and 1990s, which reduced substantially farmers' ability to gain access to inorganic nutrients by cutting subsidies and credit programmes. The fact that the World Bank is behind both structural adjustment and the SFI has sent a set of contradictory signals to African governments, who hope that the SFI may provide a vehicle for the reintroduction of such measures.

New directions? Recognizing the multiple dimensions of soil-fertility management

Interventions to address soil-fertility management need to consider five dimensions which frame the range of options available. First, as has been demonstrated throughout this book, the diversity of sites, soils and strategies found within and between African farming systems is very great. Thus, tailoring approaches to suit the opportunities and problems encountered in any particular location is of key importance. Second, there is a considerable level of differentiation between farmers in any one location, which means that no single package of measures will be appropriate for all. Third, as was clear from the discussion above, there is a wide choice of potential actions aimed at achieving better soil-fertility management, from direct technical inputs to broader macro-economic measures, which can be pursued either separately or in combination. This presents decision-makers at different levels with a valuable menu of options from which to choose. The fourth dimension concerns the timeframe over which such measures might usefully be implemented, given current conditions and other ongoing policy changes with an influence on soil management. Some measures produce rapid impacts, particularly those operating through prices and markets, while others are far slower in bringing about changes such as land tenure reforms, and changes to training of extension officers. The fifth dimension concerns the broader context, and how macro-level decisions might better take into account the need to promote more sustainable patterns of soil management at farm level. Hence the potential role to be played by a national strategy for soil-fertility management which forces an explicit analysis of how current policies in a range of fields affect incentives at farm level.

A sustainable-livelihoods approach to policy design

So how can these multiple dimensions of soil-fertility management be combined? As the case study research has shown, an exclusive focus on the technical aspects of soil-fertility management is unlikely to generate effective interventions and appropriate policies. Soil-fertility management is but one part of a broader set of livelihood activities that rural people pursue. Taking a more holistic analysis, then, a sustainable rural livelihoods approach outlines the range of different 'capital' assets which such interventions might address, as discussed in Box 6.1. As can be seen, there are many routes to achieving the goal of improved soils management.

For each of the five forms of 'capital' described in Box 6.1, there will be a range of intervention options to be considered, depending on the assets and livelihood strategies of different groups of people, with men, women, younger people, older people, in-migrants and others requiring different forms of support. Thus, for example, ways of improving natural capital will need to consider how to address the particular constraints of poorer farmers with small fields and poor soils, who also have limited access to cash and labour. Thus, attention might usefully be paid to forms of micro-credit best suited to poorer groups, as well as getting extension staff to focus on the differentiated needs of farmers, instead of targeting the 'average' farm household. By contrast, the options open to better-off farmers in a given site are likely to be much broader and, therefore, prompt a different set of interventions.

The way different social groups gain access to these different forms of 'capital' is dependent on a complex interaction of institutions and organiza-

BOX 6.1 RURAL LIVELIHOODS: IDENTIFYING AVENUES FOR INTERVENTION FOR SOIL-FERTILITY MANAGEMENT

- *Natural capital*, through a range of direct interventions aimed at improving the biophysical status of soils, such as recapitalization, use of chemical inputs and build-up of organic matter, and construction of anti-erosion measures.
- *Financial capital*, by support to credit and savings schemes to facilitate the import of nutrients onto farms and investment in labour, transport, or livestock for their manure.
- *Physical capital*, by building roads and other means of communication to improve access to markets and thus shift relative prices and improve incentives for soil-fertility management.
- *Human capital*, through working with and building on farmers' knowledge and skills, to develop more effective partnerships between farmers, research and extension staff.
- *Social capital*, by improving the organizational capacity of farmers to work together, experiment with alternative technologies, reflect and evaluate options and identify needs from technical service agencies.

Source: Scoones and Toulmin, 1999

tions operating at different levels. The institutional process which mediates access to such assets is therefore a critical component of any livelihoods analysis. For example, in Ethiopia, depending on social status, religious affiliation and the ability to pay in kind for labour, different people may gain access to labour for soil management and other agricultural activities through a variety of different local institutional forms (see Chapter 2). Such local institutions may interact with more formal organizational structures of the state or development agencies in various ways. Thus, for example, local, largely informal credit or savings organizations operating among networks of kin, friends or church members may be usefully enhanced by externally-supported credit arrangements to improve the ability of members to increase financial capital. It is this interlocking, multi-layered nature of informal and formal organizational arrangements which is central to an understanding of how livelihoods are constructed by different people in different settings. Insights into the changing nature of institutional configurations across levels, then, can assist with the sensitive design of appropriate interventions which enhance people's own capacities to manage resources.

Combining different approaches

Much of the current international debate on agriculture in Africa pays particular attention to direct interventions aimed at increasing soil nutrients, through supplementing natural capital by, for example, increasing use of inorganic fertilizer. However, there may be more effective means of achieving the objective of improving livelihood sustainability. Box 6.1 suggests other avenues which should also be considered in identifying a broader spread of options for tailoring to local conditions.

In any particular setting, it is unlikely that a single intervention, by itself, will make a big difference to soil-fertility management and improved livelihoods. For example, in agricultural systems where soil organic matter is in short supply, substantial inputs of N may be needed initially to generate sufficient biomass to contribute to the longer-term objective of improving soil structure and building organic matter content. At the same time, work could be initiated through farmer field schools and experimentation aimed at increasing human and social capital through raising skills, knowledge of more effective biomass management and strengthening partnerships between farmers and extension systems (see Chapter 5). At the macro-level, debate could be initiated on reforms to research and extension systems, ways of improving access to markets and credit, and strengthening of tenure security.

From the case study sites, a range of combined actions can be identified aimed at improving soil management and livelihood sustainability, which address the particular characteristics of location and farm household, as shown in Box 6.2 below.

From these examples based on the field research, a varied choice of interventions can be identified to support more effective soil-fertility management. Some of these are focused at farm and field level, while others relate to wider policy and macro-level changes. An analysis that cuts across scales – from the micro to the macro – and across intervention areas – from technical to institutional – is essential if the appropriate mix is to be found. As we have seen, however, much of emphasis in the past – and persisting to the present – has focused on technical options derived from a limited view of soil management emanating from soil and agronomic studies. The emphasis on inorganic chemical fertilizers is perhaps the dominant example of such an approach, although it could be argued that the low external input and organic farming perspective suffers from a similar weakness and narrowness of vision.

Any technical options, then, need to be seen in a broader context. This points to the importance of a wider socio-economic and institutional analysis, asking who gains and who loses from different options? How are different intervention options linked into broader patterns of livelihood change? Which combination of institutional and policy factors operate and interact at different levels, and how do they help direct farmers along more desirable pathways of change?

Strategies for Integrated Soil-Fertility Management: Following a Phased Approach

As previous sections have shown, a long-term strategy for integrated soil-fertility management needs to take account of a wide range of factors, from the macro to the micro, and across a huge range of policy areas. Such a strategy needs to consider how to implement such a range of measures over time, and best link local-level practice and national level policy. This would allow for the design of a set of interventions to be implemented, which are tailored to particular settings and able to adapt to changing circumstances.[3] A phased strategy

BOX 6.2 IDENTIFYING OPTIONS FOR POLICY AND PRACTICE: EXAMPLES FROM THE FIELD SITES

Mali

In the irrigated rice village of Tissana in Mali, the diversification by farmers into crops other than rice is set to continue. Further growth of the scheme is also underway, with investment in construction of new canals and irrigation works into neighbouring dryland areas. Interventions to improve soil-fertility management within the Office du Niger thus need to include a combination of:

- A shift in research and extension from an exclusive focus on irrigated rice, grown with inorganic fertilizer, towards support for other crops, combining organic and chemical materials to best effect.
- Work with newly-settled rice farmers who are less familiar with inputs and credit, water management, and marketing, as well as how best to maintain and improve the quality of soil on their plots, combined with attention to areas long under cultivation where soils are suffering from salinization and loss of structure.
- Seeking maximum complementation between livestock and cropping enterprises in and around the Office du Niger. Cattle herds depend on gaining access to dry-season grazing, while at the same time providing manure which is becoming a key component of the farming system.

Ethiopia

In the highland site in southern Ethiopia, livelihood options are increasingly constrained. Limited land areas and the lack of available oxen and cash to invest in increasing agricultural production on the small plots of land means that a flexible strategy involving a range of farm and non-farm activities is essential. The current focus for rural development in this area concentrates on an extension package involving improved seeds and fertilizer linked to a credit arrangement, which is implemented in an inflexible manner. Only the better-off can risk the package and gain the undoubted benefits. By contrast, this option increases vulnerability for poorer farmers, as they have no other sources of income or land to fall back on in case of failure. Alternative options could include:

- A wider set of technical options as part of the package, which involve lower-risk crops and management practices.
- A more flexible form of package so that farmers may take different elements of the package in relation to the condition of their soils, their production objectives, their asset status and risk preferences.
- An alternative form of 'credit for livelihoods' which would allow the allocation of credit to a range of livelihood activities to encourage agricultural improvement in the context of wider livelihood diversification.
- The linking of formal credit (from government and private sources) to informal institutions governing savings and mutual assurance.

Zimbabwe

In Zimbabwe, the increasingly uneconomic option of applying inorganic fertilizer at recommended rates has resulted in declining use, even in higher-potential areas where agronomic responses are relatively good. This has potentially negative impacts on the viability of agriculture in these areas, with consequences for economic growth and food security. Farmers' own experiments with combinations of organic and inorganic amendments have not received much attention from research and are not recognized in standard extension support. A range of options for improving the situation suggest themselves.

- A revision of fertilizer recommendations to take greater account of local agro-ecological setting and socio-economic circumstance, with a set of graded options with different mixes offered by local traders.
- Support for new fertilizer traders and suppliers to encourage a growth in the market. A variety of public–private partnerships could be explored which might help reduce trading margins, particularly in more remote areas, and so reduce farm-gate costs.
- Different bagging options and more variety in fertilizer mixes, with advice offered by retailers, would allow greater customer choice, and perhaps greater demand, particularly from those requiring relatively small amounts for specific uses.
- More emphasis in research and extension on improving fertilizer-use efficiency through different placement and timing options. Combined with innovative ways of combining inorganic with organic sources, limited supplies of fertility inputs could be made to go further, thus increasing productivity.

would therefore need to start with a period of participatory planning and assessment, the development of new skills, and be linked to a series of decisions regarding the appropriate scale and vehicle for intervention.

A first step would be an initial *assessment of context and constraints* to identify the biophysical, socio-economic and institutional characteristics of the district, province or commune where work is planned. In parallel, a macro-level analysis is needed of policies which affect the pattern of incentives for farmers to manage their soils more sustainably, given the range of other opportunities and constraints they face, in the agricultural sector and elsewhere. At farm and community levels, *participatory planning and analysis* enables local people, researchers and extension staff to identify a set of activities which farmers want to try, and to establish methods for joint reflection and evaluation. Experience with participatory extension approaches in combination with farmer field schools provides a variety of practical tools for supporting farmers in the analysis of problems, choice of options to test out, and strengthening of organizational links to help spread ideas and discussion amongst farmers (see Chapter 5). At the same time, an *assessment of organizational setting* is needed to identify the current strengths and capacity of different channels through which a combination of interventions might be supported. This will depend on existing structures within the existing governmental, NGO and private sectors, the skills and resources available, and the flexibility and openness of different structures to working with farmers in a more intensive and collegiate manner.

As already discussed, the tailoring of soil-fertility interventions to the diversity and dynamics of particular contexts will need to go beyond a purely technical focus, to embrace a much wider set of options (see Boxes 6.1 and 6.2). The *skills* needed by research, extension and development agents to take forward the approach outlined above will need to include:

- Economic and social analysis to understand the diverse constraints faced by farmers and the historical dimension to the farming system's development, in order to set the particular issue of soil-fertility management within the broader context of farming livelihoods.
- Participatory planning, analysis and facilitation of farmer-led experimentation, by support to processes of learning and exchange, and major changes in the roles of research and extension staff. Such changes are partly underway in many research and extension structures but require further commitment to ensure their firmer establishment, and the integration of participatory approaches within the way such organizations operate.
- Institutional and organizational analysis to identify structures with which to work and pathways along which the goal of improved soil-fertility management can be achieved. For example, in Mali and Zimbabwe there is a much greater role now being played by the private sector (traders, transporters, shopkeepers) in the provision of inputs and purchase of crops. Organizational analysis needs to assess, for example, how best to build on the energy and flexibility afforded by the local private sector to improve access to inputs and markets.

However, it takes a long time to change the ethos and skills-base available to institutions such as government research and extension agencies. Hence, a long-term programme for training and re-training may be required, as well as reliance in the short term on other sources of expertise, such as the NGO community.

Developing a strategy for integrated soil-fertility management will also mean making choices about level and strategy. Trade-offs and synergies must be assessed between the following strategies:

- *A local-level focus*, based on a participatory learning approach which gradually builds capacity at this level through the development of skills, pilot projects to test out methods of working with farmers, training of trainers and methods of spreading experience with possible partners. There may already exist a body of organizations with considerable experience in this field on which to build. Even if lacking within a given country, there may be useful experience in neighbouring countries on whose skills such a locally-focused programme could be based. Support to networking amongst the various organizations working on participatory soil-fertility management could be one among several ways of spreading such approaches, through exchange of experience and lesson-learning. Attention must also be paid to structures of incentives faced by researchers and extension agents to adopt new methods of work, other changes underway in the agricultural research and NGO worlds, the space for local fora to be established and opportunities for linking debate at local level with higher levels.

- *A centralized approach with a technical focus*, such as distribution of rock phosphate supplies for recapitalization of soils. This would require a well-planned system for organization of delivery, instructions to farmers regarding its use and methods for recouping costs. The newly-completed National Strategy for Soil-fertility Management for Burkina Faso provides an example of such a technical focus involving the distribution of substantial volumes of rock phosphate mined in the north of the country to farmers in the centre and north-west (Government of Burkina Faso, 1999). The cost of this activity, estimated to be 5.9 billion CFA francs (equivalent to US$9.7 million), may well be covered from a range of interested donors who often find it easier to provide a single large payment of funds than a smaller, regular commitment.
- *A macro-policy focus*, which aims to influence policy measures in ways likely to improve the incentives faced by smallholder farmers. Such a macro-level focus requires that farmers be sufficiently well-integrated into economic and policy circuits for changes at macro-level to have an impact at farm level. This is much more assured where farmers are already engaged in cash-crop production and reliant on significant levels of purchased inputs. In such cases, price changes for inputs and outputs can have a major and rapid impact on choice of crop, and the level and type of soil-fertility inputs used. In other areas more distant from markets, pricing policy will have more muted effects. Use of macro-level instruments also assumes a willingness in government policy and donor circles to develop a more coherent approach to addressing soil-fertility management for improved livelihoods. Without such coherence, the incentives faced by farmers are, in practice, merely the net result of decisions made in a number of areas with no explicit account taken of their impact on soil management.

As part of any consideration of phasing external support, it may be appropriate to look at how such approaches may be combined at different points. It may, for instance, be essential to address broader macro-policy issues before embarking on a local-level approach, as without such an enabling context at higher level, local initiatives may fail. Equally, it may be appropriate to aim for a top–down, technical intervention (such as fertilizer supply) to address the immediate consequences of short-term food insecurity and build towards a more long-term integrated and participatory approach over some years.

FUTURE DIRECTIONS?

The approach outlined above contrasts in some important respects with the current round of environmental strategies and conventions which are in the process of preparation and implementation. These include the UN Convention to Combat Desertification, the Soil-fertility Initiative and

National Environment Action Plans. One obvious problem raised by these environmental strategies is the degree of overlap, duplication and waste involved in pursing often similar objectives but through different structures, which actively compete with each other, both at national and international level. A second issue concerns the very high cost of preparing such global strategies in comparison with the funding available to implement what has been agreed. For example, it is reckoned that the cost of negotiating the UN Convention to Combat Desertification must far exceed US$100 million over the period from May 1993 (Toulmin, 1997). It is reasonable to ask whether such sums are justifiable given the end product and the likelihood of the intended beneficiaries gaining anything tangible from the National Action Programmes currently being formulated. But the third, and most important, problem concerns the approach taken by these international plans and strategies. Despite a rhetorical flourish in favour of participation and consultation, and a nod in recognition of diversity, such initiatives continue to be led by international donors, and emphasize the important role of international coordination and facilitation, rather than taking measures to ensure real interest and ownership in the countries concerned.

It is vital to have an agreed consensus regarding the nature of the policy problem being faced, and the system of which it forms part. The 'crisis narrative' expressed in many international statements about the state of African agriculture tells a particular story which is appealing in its simplicity. It suggests that things are bad and getting worse, and something must be done urgently. While this may be good at raising the interest of donors and others, it may result in poorly thought-out and hastily-implemented projects which make it difficult for other more considered approaches to be carried out. Such interventions may, in some cases, undermine livelihoods and reduce the potential for sustainability in the long term. Programmes which stress inorganic fertilizer use may, for example, push out a more balanced approach including biomass management which would help assure the longer-term structure and productivity of soils.

An alternative, more cautious way forward is needed. Such an approach does not state that 'there is no soil-fertility problem', but rather that problems are local, specific, differentiated and dynamic and will require mostly local efforts to be addressed effectively. This approach – elements of which have been outlined in different parts of this book – proceeds through a combination of farmer experimentation, monitoring and sequential learning as part of a longer-term participatory process. It is less glamorous than global initiatives and spends aid money less on technical aspects and more on building local skills and reforming institutional and policy processes. Such an approach pays particular attention to the phasing and skills required for pursuing this kind of programme, it acknowledges the very diverse set of conditions and practices found at farm level, and it recognizes that African farmers have been very creative in their ability to adapt and cope with rapid changes to the economic, institutional, technical and political settings in which they find themselves. While such an approach focuses on the local and the particular, it recognizes the importance of the broader macro conditions within which local practices are set.

Those in favour of this approach need to establish a rather different kind of debate at global level which avoids simplification. The narrative for such an approach tells a story of diversity and dynamics and the need to support a set of locally-generated processes, pay attention to institutional and policy settings and place soil-fertility management issues in their broader livelihood context. It is hoped that this book can make a contribution to this alternative vision and so help set the debate on how best to support soil management in Africa on a new and more productive path.

NOTES

CHAPTER 1 TRANSFORMING SOILS

1 This section draws on *Policies for Soil Fertility Management in Africa* by Scoones, I and Toulmin, C (1999) DFID, London
2 These are derived from a range of the major players in the international scene, including the World Bank (eg World Bank–FAO, 1996); the UN Food and Agriculture Organization (eg FAO, 1995); the UN Environment Programme (eg UNEP, 1992); and members of the CGIAR (eg CGIAR, 1995; IFPRI, 1995; ICRAF, 1996); Keeley and Scoones, 2000b. The Soil Fertility Initiative led by the World Bank and the FAO emerged as a focus from 1996. Since then, other initiatives have included the proposal for a specific 'Soils Convention'
3 Debates about 'desertification' have a long history (see Swift, 1996), but really became a major international policy concern following the Sahelian droughts of the 1970s. The Nairobi conference of 1977 was key in setting the parameters of the debate, and this has been followed by a variety of initiatives by UNEP and others leading up to the signing of the Convention to Combat Desertification in 1994, following on from the UNCED conference of 1992 (Toulmin, 1993)
4 A changing focus on research is reflected in the profile of work published in key regional agricultural journals. For example, an analysis of published articles in both the *East African Agriculture and Forestry Journal* and the *Rhodesia Agricultural Journal* (later *Zimbabwe Agricultural Journal*) shows how the 1930s in particular were dominated by discussion of soil conservation issues and a range of organic matter management techniques, notably green manuring, composting and mulching. For example, in the *RAJ* some 30 articles were published on composting and green manuring during the 1930s, while only 13 were published in the subsequent 23 years. During the 1940s a concern with rotations and ley systems is evident, particularly in East Africa, although this had a longer tradition of research in Rhodesia dating back to the 1910s. From the 1950s papers on soil fertility issues are increasingly dominated by fertilizer experiments. For example, in the *EAFJ* only 17 articles were on inorganic fertilizers in the 20-year period from 1936 to 1955, while in the following 20 years to 1975, 30 articles were published on the subject, with another 22 published in the decade that followed
5 Soil classifications took a variety of forms. Sometimes they remained very generic (differentiating only sands, clays, loams etc); on other occasions they adopted schema being more widely used around the world, whether the USDA soil taxonomy, the FAO–UNESCO legend, or the ORSTOM system (Young, 1976; Dalal-Clayton, 1990). In many cases, specialized soil classifications suited to local conditions were developed, as in parts of southern Africa (eg Federation of Rhodesia and Nyasaland, 1962) or West Africa (Ahn, 1970). To this day, despite attempts by international organizations such as the FAO to settle for a single format, there remains no global standard (although see FAO, 1998a)

6 For example, during the 1930s more detailed soil surveys were carried out in Sierra Leone (1932), Malawi (1938), Zaire (1938), Kenya and Tanzania (1935–1936) (Young, 1976; Russell, 1988; Dalal-Clayton, 1990). In some places, for example in northern Zambia, such soil surveys formed the central component of a wider regional analysis of farming systems (Trapnell and Clothier, 1937; Trapnell, 1943)
7 The long-term trials established in Harare, Zimbabwe (from 1913), Serere, Uganda (from 1937), Kabete, Kenya (from 1976), Ibadan and Samaru, Nigeria (from 1950), Saria, Burkina Faso (from 1960) were all part of this widespread research endeavour
8 For example, the FAO, through its 'Freedom from Hunger' campaign, launched a major programme of fertilizer testing across Africa in 1960 (FAO, 1958 and 1960)
9 Indeed, the FAO, long the champion of the chemical fertilizer approach, recently renamed its fertilizer section to reflect a new focus on more integrated approach. Integrated nutrient management (see Janssen, 1993; Palm et al, 1997) emphasizes, in particular, the relative balance of a variety of different inputs (including mineral fertilizers) against the range of outputs from a farm or plot. The FAO-commissioned work by Stoorvogel and Smaling (1990) was highly influential in providing a methodology to reinforce a shift to a more integrated perspective
10 It is notable that these two themes first appear as a recurrent focus for publication in the *EAFJ* from the 1980s
11 In recent years a growing interest in sustainability issues has focused attention on broader issues of soil management. The work by the Tropical Soil Biology and Fertility (TSBF) programme based in Nairobi has been particularly influential, as has support from the Rockefeller Foundation in southern Africa. This more technical scientific work has complemented other initiatives emerging out of the land husbandry, organic farming and sustainable agriculture approaches promoted by such organizations as the Association for Better Land Husbandry (ABLH) and the Centre for Research and Information on Low-External-Input and Sustainable Agriculture (ILEIA)
12 See Ingram, 1994; Morse, 1996; Gregory and Harris, 1997; Sanchez et al, 1997; Pieri, 1989; and Buresh and Giller, 1998 for more extensive reviews
13 A wide range of data sources and methodologies have been used for the data presented in this table, with the result that the figures are not exactly comparable. For example, the positive figures for average balances in southern Mali are the result of significant nutrient inputs from inorganic fertilizer, and nutrient transfers through grazing of animals on common pastures. The table should therefore only be used to assess broad patterns and the specific studies should be consulted for methodologies. An important source is Smaling (ed) (1998) where many of the cited case studies are reported
14 Studies from different African settings have calculated (with a wide variety of assumptions) that the amount of rangeland required to sustain a hectare of arable land through manure transfers ranges from 5ha (McIntire and Powell, 1995) to 20ha (Fernandez-Rivera et al, 1995; Breman et al, 1990) (see Scoones and Toulmin, 1999 and Turner, 1995 for futher discussions of these studies)
15 While the increasing interest in farmer participatory research since the 1980s has reversed this trend somewhat, budget constraints in most national agricultural research systems have limited the level of on-farm work severely
16 Environmental economics assessments are perhaps the only exception (for example, Bojö (1991, 1996), Bishop and Allen (1989)) where the social sciences have contributed to the soil degradation debate
17 For example, Defoer and Kanté (1996) for Mali; Scoones (1997) for Zimbabwe; Östberg (1995) for Tanzania; Fairhead and Leach (1996) for Guinea; Richards (1985) and Nyerges (1997) for Sierra Leone. See Chapters 2–4 for further attempts

18 For example, attempts were made during Farming Systems work from the 1980s, particularly through the influence of the CIMMYT Economics Programme which encouraged basic economic analysis of input options

19 An adaptation of the definition proposed by Abel and Blaikie (1989) for rangeland settings would be 'an effectively permanent decline in the rate at which land yields products useful to local livelihoods within a reasonable time-frame'. Such a definition focuses on the human use of the environment, rather than any inherent biophysical changes. It emphasizes permanent shifts which cannot be reversed by human investment, and therefore is not concerned with transient shifts in productivity levels due to rainfall or other factors (see Scoones and Toulmin, 1999, for a further discussion)

20 For more detail on the methodologies used in this and other similar work, see Defoer and Budelman (2000)

21 Structural adjustment programmes were implemented in Mali from 1982. In Zimbabwe, the first phase started later in 1991. In Ethiopia some early measures were applied in 1991 following the fall of the Derg, with the pace of policy reform picking up in the following few years

22 With the spread of the HIV/AIDS pandemic, population growth rate projections are dramatically decreasing, particularly in southern Africa

23 This is common pattern across Africa (eg Prudencio, 1993; Fairhead and Leach, 1996)

24 Many exceptions of course exist to this generalization, especially in west Africa where well established patterns of female farming exist (see Guyer, 1984)

25 A number of important precursors to this work however exist. Work by Östberg (1995), Sillitoe (1996) and Fairhead and Leach (1996) are all good examples of a more socio-ecological perspective to understanding soils and land management

26 See, for example, work commissioned by GTZ (Steiner, 1996) and the World Bank (Pieri et al, 1995) which suffers some of these problems

27 A 'state and transition' approach to analysing environmental change has its origins in attempts to understand highly dynamic rangeland settings (Westoby et al, 1989; Walker, 1993; Behnke and Scoones, 1993). In some rangeland situations, high levels of variation in the biotic environment (eg in dryland areas where rainfall variation between seasons and years is hugely unpredictable and variable) result in a non-equilibrium form of dynamics where shifts between states are uncertain and contingent. In other situations, biotic and abiotic factors interact over different time scales to effect change and a series of partial equilibrial states can be identified. Transitions between states, particularly in the partial equilibrium settings, may be influenced by a range of management choices. These require an understanding of the socio-economic and policy factors that influence the frequency of transitions. Most state and transition approaches as applied to grasslands, however, have concentrated on ecological questions. The approach discussed here develops this further to incorporate broader institutional factors which mediate change (see Kepe and Scoones, 1999, for an example from a South African grassland setting)

28 See Carney, 1998 and Scoones, 1998 for a discussion of approaches to analysing sustainable livelihoods using an asset-based framework

29 See for example, Turner et al, 1993; Tiffen et al, 1994; Mortimore, 1998

30 Applications of inorganic fertilizer in the cotton zone increased following the devaluation of the CFA (the West African franc) due to favourable price conditions. A switch in applications to cereals, notably maize, also occurred, particularly following the decline in cotton prices

CHAPTER 2 CREATING GARDENS

1 The team included: Alemayehu Konde, Data Dea, Ejigu Jonfa, Eyasu Elias, Fanuel Folla, Kelsa Kena, Ian Scoones, Tesfaye Berhanu and Worku Tessema. This chapter has been written by Ian Scoones on the basis of a range of field reports produced for this project, as well as other secondary materials. Research assistance was provided by Mestawot Taye and Abera Altaye
2 See for example, FARM-Africa (1992a,b)
3 The EHRS claimed gross soil losses from crop land of 130t/ha/year with major productivity collapses in the agricultural system predicted. These figures held great sway in the policy debates of this period, but have subsequently been criticized and revised downwards substantially (see Sutcliffe, 1993; Bojö and Cassells, 1995; Keeley and Scoones, 2000a)
4 A range of longer-term ways of gaining access to livestock exist, including shared ownership (*kotta*, *ulo-kotta*), share-rearing (*hara*) and profit-sharing arrangements (*tirf yegera*). These are complemented by short-term arrangements including pairing (*gatha*), borrowing for free (*woosa*) and in exchange for labour or land (Carswell et al, 2000)
5 Research on the formal recommendations emerging from the SCRP and adopted by the Ministry of Agriculture has cast doubt on their appropriateness. This has reflected in peasants destroying bunds during the change of regime in 1991 (Dessalegn, 1994b). First, questions have been raised about the amount of soil loss being experienced (Sutcliffe, 1993) and the likely returns on the investment of mechanical soil-conservation structures. Some studies show that the standard recommendations result in negative economic returns even over a long period (Herweg, 1992). This has prompted greater interest in indigenous soil and water conservation measures in Ethiopia (Krüger et al, 1996, 1997)
6 Tsetse infestation rose in the Bele area from 1 per cent in 1973 to 26.6 per cent in 1993 (data from FARM Africa survey and review)
7 Similar surveys show comparable patterns of socio-economic differentiation in Wolayta. See, for example, surveys by Dagnew, 1993, 1995; SOS-Sahel, 1993; Dessalegn, 1990; Carswell et al, 2000
8 For example, FAO (1995) points to 'alarming' rates of nutrient loss at a national level. Calculated from a generalized model they estimate that 157,000 tonnes of N is lost through cropping each year. This excludes the additional losses from erosion, which are also estimated to be considerable if national level erosion figures are used to calculate nutrient loss (Sutcliffe, 1993)

CHAPTER 3 SEIZING NEW OPPORTUNITIES

1 The team included: Ibrahim Dembéle, Loes Kater, Daouda Kone Yenizie Koné, Boutout Ly and Allaye Macinanke. This chapter has been produced with additional inputs from Arnoud Budelman, Toon Defoer, Thea Hilhorst, Ian Scoones and Camilla Toulmin
2 An approach which emphasizes the planning and management of land and other natural resources within the territory surrounding a village
2 The Inter-governmental Panel on Climate Change (IPCC) estimates that overall rainfall levels in the Sahel may rise with global warming, but that higher temperatures will lead to increased rates of evaporation. As a result, soil moisture may well fall. In addition, the IPCC reckons that increased rainfall will be accompanied by heavier and more violent storms, bringing increased run-off and erosion. Thus, ways need to be sought to help farmers improve soil conservation and moisture retention, to reduce risks of erosion, and ensure best use of available rainfall

3 Fears of desert advance have been associated with the region since colonial times and came to the fore again with the drought years of the early 1970s. Another harsh drought in 1984 provoked further fears of continued desiccation. So far as the movement of the desert itself is concerned, recent research provides clear evidence for there being no physical shift in the desert margin due to human activity (Nicholson et al, 1998), with patterns of vegetation highly dependent on annual rainfall
4 For example, Bishop and Allen's study extrapolates plot level data from studies carried out in Burkina Faso to a national level in Mali
5 Such as the Projet Lutte Anti-Erosif, Koutiala (Hijkoop, 1991; Vlaar, 1992)
6 The Office du Niger has the advantage of relying on gravity irrigation, rather than pumps, as in Senegal. As a result, the Office in Mali has done relatively well following the 1994 CFA Franc devaluation, while in Senegal, higher pump fuel costs have placed a heavy additional burden on irrigated farming
7 Researchers in Sikasso have developed alternative recommendations to the blanket formula used in Southern Mali, which the CMDT has been experimenting with for the last two years
8 *Sègè* and *kata* are terms used to describe potash, made out of burning cereal stover, and extracting the salts, then used for various purposes, such as soap making and adding to *tô*, the local millet dish
9 In one village, Siguiné, only a single year was covered

CHAPTER 4 SOILS, LIVELIHOODS AND AGRICULTURAL CHANGE

1 The team included: C Chibudu, G Chiota, E Kandiros, B Maredzage, B Mombeshora, M Mudhenra, F Murimbarimba, A Nasasara, and I Scoones. The work was coordinated in Zimbabwe by Bright Mombeshora, team leader of the Farming Systems Research Team, and supported from Europe by Ian Scoones of IDS. This chapter has been compiled from a variety of sources and written by Ian Scoones
2 The quotes from native commissioners (NCs) are derived from district annual reports held in the Zimbabwe National Archives
3 Input levels have not increased over the period. Indeed, they probably have decreased
4 To date, fertilizer companies have not sold fertilizer in smaller amounts, although some NGOs (notably CARE in Masvingo) have been experimenting with selling on fertilizer in smaller packages through a network of rurally-based traders, with some success

CHAPTER 5 PARTICIPATORY APPROACHES TO ISFM

1 This suggests approaches that make use of Bayesian rather than normal statistics, together with qualitative monitoring of key parameters, rather than any attempt to capture everything quantitatively
2 A resource guide based on the experiences of the country teams (and others) is in preparation, aimed at providing an approach for participatory learning and action research, including adapted tools for conducting field research and resource-flow analysis for use by field practitioners (Defoer and Budelman, 2000)
3 Some important lessons can be learnt from the integrated pest management experience, where similar attempts to encourage farmer-level analysis of pest problems in order to define locally-relevant integrated solutions have been attempted (van der Fliert, 1993)

CHAPTER 6 WAYS FORWARD?

1 It is beyond the scope of this chapter to review these in detail, but many other studies provide useful overviews. For example, Bumb and Baanante (1996) and Larson (1993) on fertilizer use; Lal (1984) on soil conservation; Greenland and Nye (1959) on fallows; Vogel (1994) on tillage; Mugwira and Murwira (1997) on animal manures; Giller and Cadisch (1995) on nitrogen fixation and Jones et al (1997) on legume residue use
2 For example in Zimbabwe, 10t of manure per hectare every three years is the standard recommendation
3 See Defoer and Budelman (2000) for a set of practical guidelines about how to develop an integrated soil-fertility management approach at farm level over a series of cropping seasons

REFERENCES

Abel, N O J and Blaikie, P (1989) 'Land degradation: stocking rates and conservation policies for the communal rangelands of Botswana and Zimbabwe', *Land Degradation and Rehabilitation*, vol 1, pp101–123

ADD/NFIU (Agricultural Development Department/National Fertiliser Input Unit) (1988) *Results of NPK Trials Conducted on Major Cereal Crops*, Agricultural Development Department and National Fertilizer Inputs Unit, Ministry of Agriculture, Government of Ethiopia, Addis Ababa

ADD/NFIU (Agricultural Development Department/National Fertiliser Input Unit) (1992) 'Results of fertiliser trials conducted on major cereals, 1988–1991', *Joint Working Paper*, no 43, Agricultural Development Department and National Fertilizer Inputs Unit, Ministry of Agriculture, Addis Ababa

Ahn, P (1970) '*West African Soils*', Oxford University Press, Oxford

Akilu, B (1980) 'The diffusion of fertiliser in Ethiopia: pattern determinants and implications' *Journal of Developing Areas* vol 14, pp387–399

Alemayehu Bekele (1992) 'Fertiliser marketing in Ethiopia: past and present', paper presented at the Fifth African *Fertilizer Trade and Marketing Information Network* 10–12 November, Lome, Togo

Allen, T and Starr, T (1982) *Hierarchy Perspectives for Ecological Complexity*, Chicago, University of Chicago Press

Alvord, E (1948) 'The progress of native agriculture in Southern Rhodesia', *The New Rhodesia*, vol 15, pp18–19

Anderson, D (1984) 'Depression, dust bowl, demography and drought, the colonial state and soil conservation in East Africa during the 1930s', *African Affairs*, 83, pp321–343

Appadurai, A (1991) *The Social Life of Things*, Cambridge University Press, Cambridge

Appadurai, A (1997) *Modernity at Large, Cultural Dimensions of Globalization*, Oxford, Oxford University Press

Asfaw, N, Kisan, G, Mwangi, W and Beyene, S (1997) 'Factors affecting the adoption of maize adoption technologies in Bako Area, Ethiopia', *Ethiopian Journal of Agricultural Economics*, vol 1, pp52–73

Ashby, J, Garcia, T, Guerrero, M, Quiros, C, Roa, J and Beltran, J (1995) 'Institutionalising farmer participation in adaptive technology testing with the 'CIAL'', *Agricultural Research and Extension Network (AgREN)* Network Paper 57, London, UK, Overseas Development Institute

Ashby, J and Sperling, L (1994) 'Institutionalising participatory, client-driven research and technology development in agriculture', *AgREN paper*, 49, ODI, London

Assefa Admassie (1994) *Analysis of production efficiency and use of modern technology in crop production in Ethiopia*, University of Addis Ababa, Addis Ababa

Aylen, D (1950) 'Social and economic problems of erosion in southern Rhodesia', *Soils and Fertilizers*, vol 13(2), pp1–5

Baijukja, F and de Steenhuijsen Piters, B (1998) 'Nutrient balances and their consequences in the banana-based land use systems of Bukoba district, northwest Tanzania', *Agriculture, Ecosystems and Environment*, vol 71, pp147–158

Bationo, A, Sivakumar, M V K, Acheampong, K and Harmsen, K (1998) 'Technologies de lutte contre la dégradation des terres dans les zones soudano-sahéliennes de l'Afrique de l'Ouest', in H Breman and K Sissoko *L'Intensification Agricole au Sahel*, pp709–725, Karthala, Paris

Batterbury, S, Forsyth, T and Thompson, K (1997) 'Environmental transformations in developing countries, hybrid research and democratic policy', *Geographical Journal*, vol 163, pp126–132

Beaudoux, E and Nieuwkerk, M (1985) *Groupements Paysans d'Afrique*, L'Harmattan, Paris

Beckingham, C and Huntingford, G W B H, (trans) (1954) *Some Records of Ethiopia, 1593–1646, Series II, Vol CVII*, Hakluyt Society, London

Behnke, R and Scoones, I (1993) 'Rethinking range ecology, implications for rangeland management in Africa', pp1–30 in R Behnke, I Scoones and C Kerven (eds), *Range Ecology at Disequilibrium, New Models of Natural Variability and Pastoral Adaptation in African Savannas*, Overseas Development Institute, London

Beinart, W (1984) 'Soil erosion: conservationism and ideas about development, a southern African exploration', *Journal of Southern African Studies*, vol 11, pp52–83

Beinart, W (1989) 'The politics of colonial conservation', *Journal of Southern African Studies*, vol 15, pp143–162

Bekunda, M A, Bationo, A and Ssali, H (1997) 'Soil fertility management in Africa: a review of selected research trials', pp63–80, in R J Buresh, P A Sanchez and F Calhoun (eds), *Replenishing Soil Fertility in Africa*, Wisconsin, SSSA

Belay, T (1992) 'Erosion, its effects on properties and productivity of eutric nitosols in Gununo area' (Kindo Koisha), *Southern Ethiopia and Some Techniques of Its Control*, Institute of Geography, University of Berne, Berne

Bentley, J (1994) 'Stimulating farmer experiments in non-chemical pest control in Central America', pp147–150, in I Scoones and J Thompson (eds), *Beyond Farmer First, Rural People's Knowledge, Agricultural Research and Extension Practice*, IT Publications, London

Bereket, K and Croppenstedt, A (eds) (1995) 'The nature of share-cropping in Ethiopia, some preliminary observations', *Ethiopian Agriculture, Problems of transformation*, (ed), D Aredo and M Demeke, Proceedings of Fourth Annual Conference on the Ethiopian Economy, Addis Ababa

Bereket, K and Taddesse, M (eds), (1996) 'The Ethiopian economy, poverty and poverty alleviation', Proceedings of the Fifth Annual Conference on the Ethiopian Economy, Department of Economics, Addis Ababa

Berry, S (1989) 'Social institutions and access to resources in African agriculture', *Africa*, vol 59, pp41–55

Berry, S (1993) *No Condition is Permanent: The Social Dynamics of Agrarian Change in Sub-Saharan Africa*, University of Wisconsin Press, Madison

Bevan, P (1997) *Poverty in Ethiopia, A Background Paper*, Queen Elizabeth House, Oxford University, Oxford

Bevan, P and Bereket, K (1996) 'Measuring wealth and poverty in rural Ethiopia, a data-based discussion', pp103–124, in K Bereket and M Taddesse (eds) *The Ethiopian Economy, Poverty and Poverty Alleviation*, Proceedings of the Fifth Annual Conference on the Ethiopian Economy, Department of Economics, Addis Ababa

Bingen, R (1997) 'Technology transfer and agricultural development in West Africa, institutional challenges and opportunities in Mali', in Y Lee, *Technology Transfer and Public Policy*, Iowa State University, Greenwood Press

Birch, H and Hamito, D (1979) 'The fertility status of Ethiopian soils', paper presented to the Third Conference on Soil Fertility and Fertilizer Use in Africa, ECA, Addis Ababa

Bishop, J and Allen, J (1989) 'The economics of soil degradation', *Discussion Paper* 95–02, Environmental Economics Programme, IIED, London

Blackburn, J and Holland, J (1998) *Who Changes? Institutionalising Participation in Development,* IT Publications, London

Blackmore, A, Mentis, M and Scholes, R (1990) 'The origin and extent of nutrient enriched patches within a nutrient poor savanna in South Africa', *Journal of Biogeography*, 17, pp463–70

Bojö, J (1991) *The Economics of Land Degradation, Theory and Applications to Lesotho*, The Stockholm School of Economics, Stockholm

Bojö, J (1996) 'The costs of land degradation in sub-Saharan Africa', *Ecological Economics*, vol 16, pp161–173

Bojö, J and Cassells, D (1995) *Land Degradation and Rehabilitation in Ethiopia, A Reassessment*, World Bank, Washington

Borlaug, N E and Dowswell, C R (1994) *Feeding a Human Population that Increasingly Crowds a Fragile Planet*, Wageningen, ISSS

Boserup, E (1965) *The Conditions of Agricultural Growth, The Economics of Agrarian Change under Population Pressure*, London, Allen and Unwin

Bosma, R et al (1993) *Pour un Système Durable de Production, Augmenter le Bétaild: Rôle des Ruminants au Mali-Sud dans le Maintien du Taux Organique des Sols*, IER, Sikasso

Brand, J and Pfund, J (1998) 'Site and watershed-level assessment of nutrient dynamics under shifting cultivation in eastern Madagascar', *Agriculture, Ecosystems and Environment*, vol 71, pp169–184

Breman, H and Sissoko, K (1998) *L'Intensification Agricole au Sahel*, Karthala, Paris

Breman, H and Traoré, N (eds) (1987) *Analyse des Conditions de L'Elevage et Propositions des Politiques et Programmes Mali*, Club du Sahel/CILSS, OECD, Paris

Breman, H, Ketelaars, J J M H and Traoré, N (1990) 'Un remède contre le manque de terre?, Bilan des éléments nutritifs: la production primaire et l'élevage au Sahel', *Sécheresse,* vol 2, pp109–117

Brock, K and Coulibaly, N (1999) 'Sustainable rural livelihoods in Mali', *IDS Research Report No*, 35, Brighton

Brouwer, J and Bouma, J (1997) 'Soil and crop growth variability in the Sahel', *Highlights of Research* 1990–95, ICRISAT, Pantcheru and Wageningen Agricultural University, Wageningen

Brouwer, J and Powell, J M (1998) 'Increasing nutrient use efficiency in West African agriculture, the impact of micro-topography on nutrient leaching from cattle and sheep manure', *Agriculture, Ecosystems and Environment*, vol 71, pp229–239

Brouwer, J, L, Fussell, K et al (1993) 'Soil and crop growth micro-variability in the West African semi-arid tropics: a possible risk-reducing factor for subsistence farmers' *Agriculture, Ecosystems and Environment* vol 45, pp229–238

Budelman, A (ed) (1996) *Agricultural R&D at the Crossroads: Merging Systems Research and Social Actor Approaches*, Amsterdam, The Netherlands, Royal Tropical Institute

Budelman, A, Mizambwa, F C S, Stroud, A and Kileo, R O (1995) 'The application of the nutrient flow analysis in land use diagnostics:, the case of North Sukumaland, Lake Zone, Tanzania', paper presented at the IIED/FARM-Africa Workshop, November 26 to December 2, 1995, Soddo, Ethiopia

Bumb, B L and Baanante, C A (1996) *The Role of Fertilizers in Sustaining Food Security and Protecting the Environment: Trends to 2020*, International Food Policy Research Institute, Washington, DC

Buresh, R and Giller, K (1998) 'Strategies to replenish soil fertility in African smallholder agriculture', pp13–20, in S Waddington, H Murwira, J Kumwenda, D Hikwa and F Tagwira (eds) *Soil Fertility Research for Maize-Based Farming Systems in Malawi and Zimbabwe*, SoilFertNet and CIMMYT-Zimbabwe, Harare

Buresh, R J and Smithson, P C (1997) 'Building soil phosphorus capital in Africa', in R J Buresh, P A Sanchez and F Calhoun (eds) *Replenishing Soil Fertility in Africa*, SSSA, Wisconsin, pp111–150

Buresh, R J, Sanchez, P A and Calhoun, F (eds) (1997) 'Replenishing soil fertility in Africa', *Soil Science Society of America*, SSSA, Wisconsin

Bourzat, D and Pingali, P (1992) *Crop-Livestock Interaction in sub-Saharan Africa*, World Bank, Washington

Cadisch, G and Giller, K E (1997) *Driven by Nature:, Plant Litter Quality and Decomposition*, Wallingford, CAB Int

Camara, O S (1995) 'Utilisation des résidus de récolte et de fumier dans le cercle de Koutiala: Bilan des élements nutritifs et analyse économique', *Rapport PSS no* 18, Wageningen

Campbell, B M, Bradley, P et al (1997) 'Sustainability and peasant farming systems: observations from Zimbabwe ' *Agriculture and Human Values*, vol 14, pp159–168

Campbell, B, Cunliffe, R and Gambiza, J (1995) *Vegetation Structure and Small-Scale Pattern in Miombo Woodland*, Marondera, Zimbabwe, Bothalia, vol 25, pp121–126

Campbell, B, Frost, P, Kirchmann, H and Swift, M (1998) 'A survey of soil fertility management in small scale farming systems in north-eastern Zimbabwe', *Journal of Sustainable Agriculture*, vol 11, pp19–39

Campbell, B M, Swift, M J et al (1988) 'Small-scale vegetation pattern and nutrient cycling in miombo woodland', pp69–85, in J T A Verhoeven, G W Heil and M J A Werger (eds), *Vegetation Structure in Relation to Carbon and Nutrient Economy*, The Hague, SPB Academic Publishing

Carney, D (1998) 'Implementing the sustainable rural livelihoods approach', in D Carney (ed), *Sustainable Rural Livelihoods: What Contribution Can We Make?*, DFID, London

Carswell, G, Data, D, De Haan, A, Alemayehu Konde, Haileyesus Seba and Sinclair, A, (2000) 'Ethiopia Country Report' *IDS Research Report,* Brighton, IDS

Carter, S E and Murwira, H K (1994) *Soil Fertility Management and Farmers' Strategies to Exploit Agro-Ecological Diversity:, A Farm Survey from Mutoko Communal Area, Zimbabwe*, TSBF, Nairobi

Carter, S E and Murwira, H K (1995) 'Spatial variability in soil fertility management and crop response in Mutoko communal area, Zimbabwe', *Ambio*, vol 242, pp77–84

Carter, S E and van Oosterhout, S (1993) *Soil Organic Matter Management in Zimbabwe's Communal Areas:, A Review of the Literature*, TSBF, Harare

CGIAR (1995) *Renewal of the CGIAR: Sustainable Agriculture for Food Security in Developing Countries*, Ministerial-level meeting, Lucerne, Switzerland, CGIAR Secretariat, Washington

Chambers, R (1993) *Challenging the Professions: Frontiers for Rural Development*, IT Publications, London

Chambers, R (1997) *Whose Reality Counts? Putting the First Last,* IT Publications, London

Chambers, R, Pacey, A and Thrupp, L-A (1989) *Farmer First, Farmer Innovation and Agricultural Research*, IT Publications, London

Chiatti, M (1984) 'The politics of divine kingship in Wolaita (Ethiopia), 19th and 20th centuries', PhD thesis, University of Pennsylvania

Chivaura-Mususa, C and Campbell, B (1998) 'The influence of scattered *Parinaria curatellifolia* and *Acacia sieberana* trees in soil nutrients in a grassland pasture and in arable fields', pp191–199, in L Bergstrom and H Kirchman (eds), *Carbon and Nutrient Dynamics in Natural and Agricultural Ecosystems*, CAB International, Wallingford

Chuma, E and Hagmann, J (1995) 'Summary of results and experiences from on-station and on-farm testing and development of conservation tillage farming systems in semi-arid Masvingo', Zimbabwe, pp41–60, in S Twomlow, J Ellis-Jones, J Hagmann and H Loos (eds), *Soil and Water Conservation for Smallholder Farmers in Semi-Arid Zimbabwe, Transfer between Research and Extension*, Proceedings of a Technical Workshop, 3–7 April, Masvingo

Chuma, E and Hagmann, J (1997) 'Conservation tillage for semi-arid Zimbabwe, results and experiences from on station and on-farm interactive innovation in Masvingo', *Zimbabwe Science News*, vol 31, pp34–41

Cleaver, K M and Schreiber, G A (1995) *The Population, Agriculture and Environment Nexus in Sub-Saharan Africa: Reversing the Spiral*, World Bank, Washington DC

Cohen, J (1987) *Integrated Rural Development: The Ethiopian Experience and the Debates*, SIAS, Uppsala

Collinson, M (1982) 'A diagnostic survey of the south of Chivi District, Zimbabwe for adaptive research planning', *Occasional Papers, Report* 5, CIMMYT, Nairobi

Cousins, B (1992) *Managing Communal Rangeland in Zimbabwe, Experiences and Lessons*, Commonwealth Secretariat, London

Croppenstedt, A and Demeke, M (1995) *Measuring Technical Efficiency of Farmers in Arssi Province of Southern Ethiopia: A Random Coefficients Approach*, University of Addis Ababa, Addis Ababa

Croppenstedt, A and Demeke, M (1996) *Determinants of Adoption and Levels of Demand for Fertiliser for Cereal Growing Farmers in Ethiopia*, Centre for the Study of African Economies, Oxford

Croppenstedt, A and Demeke, M (1997) *An Empirical Study of Cereal Crop Production and Technical Efficiency of Private Farmers in Ethiopia: A Mixed Fixed-Random Coefficients Approach*, Centre for Study of African Economies, Oxford

Croppenstedt, A, Demeke, M and Meschi, M (1998) *Technology Adoption in the Presence of Constraints, the Case of Fertiliser Demand in Ethiopia*, paper prepared for the Ethiopia–Eritrea network meeting, June 1998, University of East Anglia, Norwich

Croppenstedt, A and Mamo, A (1996) *An Analysis of the Productivity and Technical Efficiency of Cereal growing Farmers in Ethiopia: Evidence from Three Regions*, CSAE, Oxford

Croppenstedt, A and Mulat, D (1997) 'An empirical study of cereal crop production and technical efficiency of private farmers in Ethiopia: a mixed fixed-random coefficients approach', *Applied Economics*, vol 29, pp1217–1226

Dalal-Clayton, B (1990) *A Historical Review of Soil Survey and Soil Classification in Tropical Africa*, International Institute for Environment and Development, London

Data, D (1996) 'Soil fertility management in Wolayta, Southern Ethiopia: an anthropological investigation', *FRP Technical Pamphlet*, 14, Farmers' Research Project, FARM-Africa, Addis Ababa

Data, D (1998) 'Soil fertility management in its social context: a study of local perceptions and practices in Wolaita, southern Ethiopia', *Managing Africa Soils*, no 1, IIED, London

Data, D and Scoones, I (forthcoming) 'Networks of knowledge: how farmers and scientists understand soils and their fertility: a case study from Ethiopia', *Oxford Development Studies*

de Groote, H, Kebé, D and Hilhorst, T (1995) 'Report from the seminar on rural financial systems in southern Mali', Sikasso, May 3–5, 1995

de Jager, A, Kariuku, I, Matiri, F M, Odendo, M and Wanyama, J M (1998) 'Monitoring nutrient flows and economic performance in Africa farming systems NUTMON IV linking nutrient balances and economic performance in three districts in Kenya', *Agriculture, Ecosystems and Environment*, vol 71, pp81–92

de Ridder, N and van Keulen, H (1990) 'Some aspects of the role of organic matter in sustainable intensified arable farming in the West-African semi-arid-tropics SAT ' *Fertilizer Research*, vol 26, pp299–310

de Vries, J and Prost, L (1994) 'L'adoption par les paysans de trois techniques dans le domaine de l'intégration agriculture-élevage, résultats des études de l'adoption des cultures fourragères, parc amélioré et fosse fumière dans la zone Mali-Sud', rapport de synthèse CMDT, Koutiala

Debrah, S K, Defoer, T and Bengaly, M (1998) 'Integrating farmers' knowledge: attitude and practice in the development of sustainable Striga control interventions in southern Mali', *Netherlands Journal of Agricultural Science*, vol 46, pp65–75

Defoer, T and Budelman, A (1998) 'Integrated soil fertility, management: knowledge encounters between famers and change agents', paper presented at the West and Central African Farming Systems Research Symposium 1998 in Bamako, pp22–25 September 1998, Bamako, Mali

Defoer, T and Budelman, A (eds) (2000) *Managing Soil Fertility, A Resource Guide for Participatory Learning and Action Research*, KIT Publications, Amsterdam

Defoer, T, de Groote, H, Hilhorst, T, Kanté, S and Budelman, A (1998) 'Participatory research and quantitative analysis for nutrient management in Southern Mali: a fruitful marriage?' *Agriculture, Ecosystems and Environment*, vol 71, pp215–228

Defoer, T, Budelman, A, Toulmin, C and Carter, J (1999a) *Soil Fertility Management in Africa: Resource Guide for Participatory Learning and Action Research, Part 1, Textbook*, KIT, Amsterdam

Defoer, T, Kanté, S and Sanogo, J-L (1999b) *Cotton Farming in Southern Mali, Part 2, Case Studies of Soil Fertility Management in Africa: Resource Guide for Participatory Learning and Action Research*, KIT, Amsterdam

Defoer, T, Kanté, S, Hilhorst, T and de Groote, H (1996) 'Towards more sustainable soil fertility management', *AgREN Network Paper*, no 63, ODI, London

Degnbol, T (1999) 'State bureaucracies under pressure: a study of the interaction between four extension agencies and cotton-producing farmers in the Sikasso Region, Mali', DPhil dissertation, International Development Studies, Roskilde University

Dejene, A (1993) 'The Iddir: a study of an indigenous informal financial institution in Ethiopia', *Savings and Development*, no XVII1, pp77–90

Dejene, A and Demeke, M (eds) (1995) 'Ethiopian agriculture, problems of transformation', Proceedings of Fourth Annual Conference on the Ethiopian Economy, University of Addis Ababa, Addis Ababa

Dembélé, I, de Groote, H and Hilhorst, T (1997) *Rôle de la femme dans la filière coton en zone Mali-sud*, CMDT, Bamako

Dembélé, I, et al (1998) 'Dynamics of soil fertility management in savanna Africa, Mali country report', Final Technical Report to EC DGXII, IIED and IDS, London and Brighton

Dème, Y (1998) 'Associations locales de gestion des ressources naturelles du Kelka, Mali', *Programmes Zones Arides Dossier*, no 74, IIED, London

Dercon, S and Krishnan, P (1996a) 'A consumption based measure of poverty for rural Ethiopia in 1989 and 1994', pp77–100, in K Bereket and T Mekonene (eds), *The Ethiopian Economy, Poverty and Poverty Alleviation*, Proceedings of the Fifth Annual Conference on the Ethiopian Economy, Department of Economics, Addis Ababa

Dercon, S and Krishnan, P (1996b) 'Income portfolios in rural Ethiopia and Tanzania: choices and constraints', *Journal of Development Studies*, vol 326, pp850–75

Dessalegn, R (1990) *A Resource Flow Systems Analysis of Rural Bolosso Wollaita*, Redd Barna and IDR, University of Addis Ababa, Addis Ababa

Dessalegn, R (1992) 'The dynamics of rural poverty: case studies from a district in southern Ethiopia', *Monograph Series* 2/92, Institute of Development Research, Addis Ababa

Dessalegn, R (1994a) 'The unquiet countryside, the collapse of "socialism" and rural agitation, 1990 and 1991', pp242–279, in Abebe Zegeye and S Pausewang (eds), *Ethiopia in Change: Peasantry, Nationalism and Democracy*, British Academic Press, London

Dessalegn, R (ed) (1994b) 'Land tenure and land policy in Ethiopia after the Derg', Proceedings of the Second Land Tenure Workshop, *Working Papers on Ethiopian Development*, 8, University of Trondheim

Deveze, J (1994) 'Les zones cotonnières entre développement, ajustement et dévaluation, réflexions sur le role du coton en Afrique francophone de l'ouest et du centre', *Caisse Française de Développement, Notes et Etudes*, no 53, Paris

Doumbia, M et al (1998) *Problématiques Agro-pédologiques Spécifiques des Grandes Zones Ecologiques du Mali,* MDRE, Bamako

Dureau, D, Traoré, B and Ballo, D (1994) *Evolution de la Fertilité des Sols en Culture Continue dans la Zone Mali-Sud, Système de Culture à Base de Cotonnier, N'Tarla*, Mali, MDRE and IER, Bamako

Ejigu, J, Tesfaye, B, Kelsa, K and Fanuel, F (1997) 'Participatory soil mapping and characterisation in Kindo Koisha, Southern Ethiopia', *FRP Technical Pamphlet*, 13, Awassa Research Centre/SOS Sahel/Farm Africa, Awassa

Elwell, H A (1974) *Soil Erosion Survey: The Condition of the Arable Land in the Farming Areas of Rhodesia*, Hatcliffe, Soil and Water Conservation

Elwell, H, A (1983) 'The degrading soil and water resources of the communal areas', *The Zimbabwe Science News*, vol 17, pp145–147

Elwell, H A (1985) 'An assessment of soil erosion in Zimbabwe', *The Zimbabwe Science News*, vol 19, pp27–33

Elwell, H A and Stocking, M A (1988) 'Loss of soil nutrients by sheet erosion is a major hidden farming cost', *The Zimbabwe Science News*, vol 22, pp79–82

Eshete, D (1993) 'The impact of food shortages on rural households of different income groups and their crisis coping strategies, a case study of Wolaita district in Ethiopia', DPhil thesis, University of Sussex

Eshete, D (1995a) 'Food shortages and household coping strategies by income groups: a case study of Wolaita District in Southern Ethiopia', in Dejene Aredo and Mulat Demeke, (eds) *Ethiopian Agriculture, Problems of transformation*, Proceedings of Fourth Annual Conference on the Ethiopian Economy, Addis Ababa

Eshete, D (1995b) 'Seasonal and exception food shortages their causes and socio-economic consequences: a case study of Wolaita District in southern Ethiopia', in *Food Security, Nutrition and Poverty Alleviation in Ethiopia: Problems and Prospects*, Mulat Demeke et al (eds), Proceedings of the Inaugural and First Annual Conference of the Agricultural Economics Society of Ethiopia, June 1995, Addis Ababa

Eyasu, E (1997a) *Soil Fertility Decline and Coping Strategies: The Case of Kindo Koisha, Southern Ethiopia*, University of East Anglia, Norwich

Eyasu, E (1997b) 'Soil fertility management and nutrient balance in Kindo Koysha farms: a case study of North Omo, southern Ethiopia', *FRP Technical Pamphlet*, 15, FARM Africa, Addis Ababa, Ethiopia

Eyasu, E and Scoones, I (1999) 'Perspectives on soil fertility change, a case study from southern Ethiopia', *Land Degradation and Development*, vol 10, pp195–206

Eyasu, E, Morse, S and Belshaw, D G R (1998) 'Nitrogen and phosphorus balances of Kindo Koisha farms in southern Ethiopia', *Agriculture, Ecosystems and Environment*, vol 71, pp93–114

Fairhead, J and Leach, M (1996) *Misreading the African Landscape, Society and Ecology in a Forest-Savanna Mosaic*, Cambridge University Press, Cambridge

FAO (1986) *Highlands Reclamation Study: Ethiopia*, Final Report, Volume 1, FAO, Rome

FAO (1988) *Soil Map of the World*, Rome, FAO/UNESCO

FAO (1995) *Integrated Plant Nutrition System Activities in Ethiopia*, FAO, Rome

FAO (1998a) *World Reference Base for Soil Resources*, FAO, Rome

FAO (1998b) 'Mali, Plan d'action national pour la gestion de la fertilité des sols, Preparation du vôlet fertilité des sols du PASOP', Aide-Mémoire, Rome

FARM Africa (1992a) *Report of Diagnostic Survey of Hanaze Peasant Association in Kindo Koisha Awraja*, MoA/SOS-Sahel, FARM-Africa, Awassa

FARM-Africa (1992b) *Report of Diagnostic Survey of Fagena Mata Peasant Association in Kindo Koisha Awraja,* MoA/SOS-Sahel, FARM-Africa, Awassa

FARM Africa (1996) *Nutrient Cycling and Soil Fertility Management Research in North Omo, Ethiopia, End of Year Report*, unpublished report, Farmers' Research Project, FARM Africa, Awassa

Farrington, J and Martin, A (1988) 'Farmer participation in agricultural research, a review of concepts and practices', *Agricultural Administration Unit Occasional Paper*, no 9, ODI, London, UK

Fernandez-Rivera, S, Williams, T O, Hiernaux, P and Powell, J M (1995) 'Faecal excretion by ruminants and manure availability for crop production in semi-arid West Africa', in J Powell, S Fernandez-Rivera, T Williams and C Renard (eds) *Livestock and Sustainable Nutrient Cycling in Mixed Farming Systems of Sub-Saharan Africa*, Volume II technical papers, proceedings of an international conference held in Addis Ababa, Ethiopia, 22–26 November 1993, ILCA, Addis Ababa

Floret, C and Serpantié, G (1993) 'La jachère en Afrique de l'Ouest', *Colloques et Seminaires*, ORSTOM, Paris

Floyd, B (1961) *Changing Patterns of African Land Use in Southern Rhodesia*, Rhodes Livingstone Institute, Lusaka

Fok, M (1994) 'L'évolution du système coton au Mali', *Cahiers agriculture*, vol 3, pp329–336

Folmer, E, Geurts, P and Francisco, J (1998) 'Assessment of soil fertility depletion in Mozambique', *Agriculture, Ecosystems and Environment*, vol 71, pp159–168

Forsyth, T and Leach, M, with Scoones, I (1999) *Poverty and Environment, Priorities for Research and Policy, An Overview Study*, UNDP, New York

Fresco, L O and Kroonenberg, S B (1992) 'Time and spatial scales in ecological sustainability', *Land Use Policy*, vol 9, pp155–167

Frost, P (1996) 'The ecology of miombo woodland', pp11–58, in B Campbell (ed), *The Miombo in Transition, Woodlands and Welfare in Africa,* CIFOR, Bogor, Indonesia

Frost, P, Medina, E, Menau, J C, Solbrig, O, Swift, M and Walker, B (1986) 'Responses of savannas to stress and disturbance', *Biology International*, no 10, IUBS, Paris

FSRU (1993) *Soil Fertility Management by Small Holder Farmers: A Participatory Rapid Appraisal in Chivi and Mangwende Communal areas*, FSRU/DRSS, Harare

FSRU (1996) *Soil Nutrient Cycling and Fertility Management in Zimbabwe: Narrative Report of Activities Undertaken in 1996*, unpublished report, FSRU, Ministry of Agriculture, Harare, Zimbabwe

FSRU (1998) 'Economic structural adjustment and communal area agriculture, case studies on the effects of policy changes on farmers' management of soil fertility in Mangwende and Chivi communal areas, Zimbabwe', pp446–486, in C Toulmin and I Scoones (eds), *Dynamics of Soil Fertilty Management in Savanna Africa*, Final report to EU, IIED, London

Gakou, A, Kébé, D and Traoré, A (1996) *Soil Management in Mali*, Sikasso, Ministere du Developpement Rural et de l'Environnement, Institut d'Economie Rurale, Centre Regional de la Recherche Agronomique

Gakou, A, Kébé, D, Sanogo, Z and Traoré, A (1997) 'Propositions d'élements stratégiques pour la gestion de la fertilité des sols au Mali', IFDC workshop, Ouagadougou

Geertz, C (1968) *Agricultural Involution: The Process of Ecological Change in Indonesia*, University of California, Berkeley and Los Angeles

Giller, K E and Cadisch, G (1995) 'Future benefits from biologial nitrogen fixation: an ecological approach to agriculture', *Plant Soil*, vol 174, pp255-277

Giller, K E, Cadisch, G, Ehaliotis, C, Adams, E, Sakala, W and Mafongoya, P (1997) 'Building soil nitrogen capital in Africa', in R, J, Buresh, P, A, Sanchez and F, Calhoun (eds), *Replenishing Soil Fertility in Africa,* SSSA, Wisconsin

Giller, K, Gilbert, R, Mugwira, L, Muza, L, Patel, B and Waddington, S (1998) 'Practical approaches to soil organic matter management for smallholder maize production in southern Africa', pp139–153, in S, Waddington, S, Murwira, J, Kumwenda, D, Hikwa and F, Tagwira (eds), *Soil Fertility Research for Maize-Based Farming Systems in Malawi and Zimbabwe*, Proceedings of the SoiFertNet Results and Planning Workshop, 7–11 July 1997, SoilFertNet and CIMMYT-Zimbabwe, Harare

Giraudy, F and Niang, M, (1996) *Impact de la devaluation sur les systemes de production et les revenus paysans en zone Mali-Sud*, CMDT, Bamako

Government of Burkina Faso (1999) *Strategie nationale et plan d'action de gestion integree de la fertilite des sols*, Ministere de l'Agriculture, Ouagadougou

Government of Mali (1998) *National Environmental Action Plan and CCD Program*, Abstract, Ministry of the Environment, Bamako

Grant, P (1967a) 'The fertility of sandveld soil under continuous cultivation: the effect of manure and nitrogen fertilizers on the nitrogen status of the soil', *Rhodesia Zambia Malawi Journal of Agricultural Research*, vol 5, pp71–79

Grant, P (1967b) 'The organic matter content of soil size-fractions', *Rhodesia Zambia Malawi Journal of Agricultural Research*, vol 5, pp211–213

Grant, P (1970) 'Restoration of productivity of depleted sands', *Rhodesia Agriculture Journal*, vol 67, pp131–137

Grant, P (1976) 'Peasant farming on infertile sands', *Rhodesia Science News*, vol 10, pp282–284

Grant, P (1981) 'The fertilization of sandy soils in peasant agriculture', *Zimbabwe Agriculture Journal*, vol 78, pp169–175

Greenland, D J (1994) 'Long-term cropping experiments in developing countries: the need, the history and the future', pp187–209, in R A Leigh and A E Johnston (eds), *Long-term Experiments in Agricultural and Ecological Sciences*, CAB International, Wallingford

Greenland, D J, Wild, A and Adams, D (1992) 'Organic matter dynamics in soils of the tropics: from myth to complex reality', in *Myths and Science of Soils of the Tropics*, SSA Special Publication, no 29, SSSA, Wisconsin

Gregory, P and Harris, P (1997) 'Managing soil fertility', paper presented at the Natural Resource Advisors' Conference, Department for International Development, London

GTZ (1995) *Causes of Soil Degradation and Development Approaches to Sustainable Soil Management*, Magraf Verlag, Weikersheim

Gunten, A (1993) 'Soil erosion process in a twin catchment set-up in Gununo area', Kindo Koisha, Ethiopia, *Research Report*, no 23, Soil Conservation Research Project, University of Berne, Berne

Guyer, J (1984) 'Naturalism in models of African production', *Man* NS, vol 19, pp371–388

Hagmann, J (1994) 'Lysimeter measurements of nutrient losses from a sandy soil under conventional till and ridge till', in B Jensen, P Schjonning, and K Mikkelsen (eds), *Soil Tillage for Crop Production and Protection of the Environment*, International Soil Tillage Research Organisation, Aalbourg

Hagmann, J (1995) 'State and effectiveness of the mechanical conservation systems for rill erosion control in semi-arid Masvingo', pp91–103, in S Twomlow, J Ellis-Jones, J Hagmann and H Loos (eds), *Soil and Water Conservation for Smallholder Farmers in Semi-Arid Zimbabwe: Transfer between Research and Extension*, proceedings of a technical workshop, 3–7 April, Masvingo

Hagmann, J, Chuma, E, Connolly, M and Murwira, K (1998) 'Client-driven change and institutional reform in agricultural extension: an action learning experience from Zimbabwe', *AgREN*, no 79, ODI, London

Harris, F (1996) 'Intensification of agriculture in semi-arid areas: lessons from the Kano close-settled zone', Nigeria, *IIED Gatekeeper Series* no 59, IIED, London

Harris, F (1997) *Nutrient Cycling or Soil Mining? Agropastoralism in Semi-Arid West Africa,* Final report to DFID, University of Cambridge, Cambridge

Heisey, P W and Mwangi, W (1995) 'Fertilizer and maize production in sub-Saharan Africa: use and policy options', paper presented at a workshop on *The Emerging Maize Revolution in Africa, The Role of Technology, Institutions and Policy*, 9–12 July 1995, Michigan State University, East Lansing

Herweg, K (1992) 'Major constraints to effective soil conservation: experiences in Ethiopia', in 7th ISCO Conference Proceedings *Sustainable Land Management Practices*, 27–30 September, ISCO, Sydney

Hijkoop, J, van der Poel, P and Kaya, B (1991) *Une Lutte de Longue Haleine; Aménagements Anti-Erosifs et Gestion de Terroir*, KIT, Amsterdam

Hikwa, D and Mukurumbira, L (1995) 'Highlights of previous, current and proposed soil fertility research by the Department of Research and Specialist Services DRSS in Zimbabwe', pp21–33, in S Waddington (ed), *Report on the First Meeting of the Network Working Group: Soil Fertility Research Network for Maize-Based Farming Systems in Selected Countries of Southern Africa,* The Rockefeller Foundation Southern Africa Agricultural Sciences Programme, Lilongwe, Malawi and CIMMYT Maize Programme Harare, Zimbabwe

Hikwa, D, Chibudu, C, Mugwira, L, Mukurimbira, L, Mabasa, S, Dhliwayo, D and Sithole, T (1997) 'Soil fertility management options for maize-based cropping systems in the small-holder farming areas of Zimbabwe', pp13–24, in M Mphepo, S Waddington and S Phmobeya (eds), *The Dissemination of Soil Fertility Technologies*, SoilFertNet, CIMMYT, Harare

Hilhorst, T and Coulibaly, A (1999) 'Elaborating a local convention for managing village woodlands in southern Mali', *Issue Paper No 78*, Drylands Programme IIED, London

Hoben, A (1995) 'Paradigms and politics, the cultural construction of environmental policy in Ethiopia', *World Development*, vol 23, pp1007–1021

Hodson, A W (1927) *Seven Years in Southern Abyssinia*, Fisher Unwin, London

Holling, C S (1993) 'Investing in research for sustainability', *Ecological Applications*, vol 34, pp552–555

Hudson, N (1961) 'An introduction to the mechanics of soil erosion under conditions of sub-tropical rainfall', *Proceedings of the Transactions of the Rhodesia Scientific Association*, vol 49, pp15–25

Hudson, N (1971) *Soil Conservation*, Batsford, London

Huffnagel, H (1961) *Agriculture in Ethiopia*, FAO, Rome

Hulme, M (1992) 'Rainfall changes in Africa, 1931–60 to 1961–90', *International Journal of Climatology*, vol 12, pp685–699

Hurni, H (1994) 'Methodological evolution of soil conservation research in Ethiopia', *IRDC Currents*, vol 8, pp17–21

ICRA (1987) 'Farming systems dynamics and risk in a low potential area, Chivi south, Masvingo province, Zimbabwe', *ICRA Bulletin*, no 27, ICRA, Wageningen

ICRAF (1996) *1995 Annual Report*, ICRAF, Nairobi

IER (1996) *Approche Recherche Action Gestion de la Fertilité des Sols, Mali*, unpublished report, Institut d'Economie Rurale, Niono, Mali

IFDC (1993) *Ethiopia, Fertiliser and Transport Sector Assessment*, USAID, Addis Ababa

IFPRI (1995) *A 2020 Vision for Food, Agriculture and the Environment in Sub-Saharan Africa*, IFPRI, Washington

IIED (1999) *Land Tenure and Resource Access in West Africa, Issues and Opportunities for the Next Twenty Five Years*, IIED, London

Ingram, J (1994) 'Soil fertility research in East Africa', *CAB Abstracts*, CAB International, Wallingford

Jacobson-Widding, A and van Beek, W (eds) (1990) *The Creative Communion: African Folk Models of Fertility and the Regeneration of Life*, Almqvist & Wiksell International, Stockholm

Jamin, J Y (1995) 'De la norm à la diversité, l'intensification rizicole face à la diversité paysanne dans les périmètres irrigués de l'Office du Niger', *Thèse CIRAD-SAR*, Montpellier, France

Janssen, B, H (1993) 'Integrated nutrient management: the use of organic and mineral fertilizers', in H V Reuler and W H Prins, *The Role of Plant Nutrients for Sustainable Crop Production in Sub-Saharan Africa*, Vereniging van Kunstmest Producenten, Leidschendam

Jenden, P (1994) *Cash for Work and Food Insecurity, Koisha Woreda, Wellaita: A Report on SOS-Sahel's Food Security Project, 1992–1994*, SOS-Sahel, Addis Ababa

Joldersma, R, Hilhorst, T, Diarra, S, Coulibaly, L and Vlaar, J (1996) 'Siwaa, la brousse sèche, Expérience de gestion de terroir villageois au Mali', *Bulletin 341*, KIT, Amsterdam

Jones, R B, Snapp, S S, et al (1997) 'Management of leguminous leaf residues to improve nutrient use efficiency in the sub-humid tropics', in *Driven by Nature: Plant Litter Quality and Decomposition*, G Cadisch and K E Giller (eds), CABI, Wallingford

Kanté, S and Defoer, T (1994) 'How farmers classify and manage their land, implications for research and development activities', Dryland Networks Programme *Issue Paper no 51*, IIED, London

Kébé, D, Vitale, J D, Sanders, J H and Djouara, H (1997) *Dévaluation du FCFA et Revenue des Agro-Eleveurs au Mali-Sud, une Analyse de Court et Long Terme*; Publication ESPGRN, IER-Sikasso, Mali

Keeley, J and Scoones, I (1999) 'Understanding environmental policy processes: a review', *IDS Working Paper*, no 89, IDS, Brighton

Keeley, J and Scoones, I (2000a) 'Knowledge, power and politics, the environmental policy making process in Ethiopia', *Journal of Modern African Studies*, vol 38, pp89–120

Keeley, J and Scoones, I (2000b) 'Global science, global policy: local to global policy processes for soils management in Africa', *IDS Working Paper*, no 115, IDS, Brighton

Kefala, Alemu and Sandford, S (1991) *Enset in North Omo Region*, Farm Africa, Addis Ababa

Kepe, T and Scoones, I (1999) 'Creating grasslands, social institutions and environmental change in Mkambati area, South Africa', *Human Ecology*, vol 27, p1

Khombe, C T, Dube, I A and Nyathi, P (1992) 'The effects of kraaling and stover supplementation during the dry season on body weights and manure production of Mashona steers in Zimbabwe', *African Livestock Research*, vol 1, pp18–23

Kindness, H (1994) *Household Cash Income Sources and Income Generating Activities in Welaita, North Omo*, FARM-Africa, Addis Ababa

Kindness, H and Sandford, S (1996) *The Economics of Fertilizer Use on Cereal Crops in Wolaita in Mid–1994*, Farm Africa, Addis Ababa

King, J and Campbell, B (1994) 'Soil organic matter relations in five land cover types in the Miombo region, Zimbabwe', *Forest Ecology and Management*, vol 67, pp225–239

Koné, Y (1997) *Organisation et Contraintes des Circuits D'Approvisionnement des Producteurs Ruraux en Engrais dans les Zones D'Encadrement de la CMDT et de l'Office du Niger*, IER, Niono

Krogh, L (1995) 'Field and village nutrient balances in millet cultivation in Burkina Faso, a village case study', *SEREIN Working Paper*, no 4, University of Copenhagen, Copenhagen

Krüger, H, Berhanu, F, Yohannes, G, M and Kefeni, K (1996) 'Creating an inventory of indigenous SWC measures in Ethiopia', in Reij, C et al (eds), *Sustainaing the Soil: Indigenous Soil and Water Conservation in Africa*, Earthscan, London

Krüger, H-J, Berhanu, F, Yohannes, G, M and Kefeni, K (1997) 'Inventory of indigenous soil and water conservation measures on selected sites in the Ethiopian highlands', Soil Conservation Research Programme Ethiopia, *Research Report* no 34, University of Berne, Berne

Kumwenda, J D T, Waddington, S R, Snapp, S, Jones, B and Blackie, M (1995) 'Soil fertility management research for the smallholder maize-based cropping systems of Southern Africa: a review', *Network Research Working Paper*, CIMMYT, Harare

Lal, R (1984) 'Soil erosion from tropical arable lands and its control', *Advances in Agronomy*, vol 37, pp83–248

Lal, R (1995) 'Erosion-crop productivity relationships for soils of Africa', *Soil Science Society of America Journal*, vol 59, pp661–667

Larson, B (1993) 'Fertilisers to support agricultural development in sub-Saharan Africa: what is needed and why?', Center for Economic and Policy Studies Discussion Paper, 13, Winrock International, Morilton

Lavigne Delville, P (1999) 'Harmonising formal law and customary land rights in French speaking West Africa', *Issue Paper* no 86, Drylands Programme, IIED, London

Leach, M and Mearns, R (1996) (eds) *The Lie of the Land: Challenging Received Wisdom on the African Environment,* James Currey, Oxford

Lightfoot, C, Dalsgaard, J, Bimbao, A and Fermin, F (1992) 'Farmer participatory procedures for managing and monitoring sustainable farming systems', *ICLARM Contribution*, no 892, ICLARM, Manilla, Phillipines

Lowdermilk, W C (1935) 'Civilization and soil conservation', *Rhodesian Agricultural Journal*, vol 32, pp553–7

Lynam, J K, Nandwa, S M and Smaling, E M A (1998) 'Introduction to the special issue of nutrient balances as indicators of productivity and sustainability in sub-Saharan African agriculture', *Agriculture, Ecosystems and Environment Journal*, vol 71, pp1–4

Magasa, A (1978) *Papa-commandant a jeté un grand filet devant nous: les exploités des rives du Niger 1902–1962*, Maspero, Paris

Maiga, A, S, N'Diaye, M K and Guindo, D (1998) 'Note sur l'intensification de la riziculture en zone Office du Niger', in Breman, H and Sissoko, K (eds), *L'intensification agricole au Sahel*, Karthala, Paris

Mapfumo, P, Mpepereki, S and Mafongoya, P (1998) 'Pigeonpea in Zimbabwe, a new crop with potential', pp93–98, in S Waddington, H Murwira, J Kumwenda, D Hikwa and F Tagwira (eds), *Soil Fertility Research for Maize-Based Farming Systems in Malawi and Zimbabwe*, Proceedings of the SoiFertNet Results and Planning Workshop, 7–11 July 1997, SoilFertNet and CIMMYT-Zimbabwe, Harare

Massou, A (1998) 'Coton, fin de règne?', *Jeune Afrique* no 1961, August

Maturuka, D F, Makombe, G and Low, A (1990) 'The contribution of economic analysis in developing promising research agendas: a maize nitrogen x phosphorus trial in Zimbabwe', *Farming Systems Bulletin, Eastern and Southern Africa*, vol 5, pp1–5

Mavedzenge, B, Mudziwo, C and Murimbarimba, F, (1999) 'Experiences of farmer participation in soil fertility research in southern Zimbabwe', *Soil Management in Africa Discussion Paper*, IIED Drylands Programme, London, UK

McIntire, J and Powell, J M (1995) 'African semi-arid tropical agriculture cannot grow without external inputs', in J M Powell et al (eds), *Livestock and Sustainable Nutrient Cycling in Mixed Farming Systems of Sub-Saharan Africa*, Volume II technical papers, proceedings of an International Conference held in Addis Ababa, Ethiopia, 22–26 November 1993, ILCA, Addis Ababa

McIntire, J, Bourzat, D and Pingali, P (1992) *Crop–livestock Interaction in Sub-Saharan Africa*, World Bank, Washington

Meillassoux, C (1975) *Femmes, Greniers et Capitaux*, Maspero, Paris

Menaut, J C, Barbault, R, Lavelle, P and Lepage, M (1985) 'African savannas: biological systems of humification and mineralization', in J Tothill and J Mott (eds), *Ecology and Management of the World's Savannas*, Australian Academy of Sciences, Canberra

Metelerkamp, H R R (1988) 'Review of crop research relevant to the semiarid areas of Zimbabwe', pp190–315, in proceedings of a workshop on Cropping in the Semi-arid Areas of Zimbabwe, Agritex/DR&SS/GTZ, Harare, Zimbabwe

Morse, K (1996) *A Review of Soil and Water Management Research in Semi-Arid Areas of Southern and Eastern Africa*, Natural Resources Institute, Chatham

Mortimore, M (1998) *Roots in the African Dust: Sustaining the Sub-Saharan Drylands*, Cambridge University Press, Cambridge

Mudhara, M (1991) *Economics and Fertiliser Use in Maize in Semi-Arid Environments:, Experiences of the Farming Systems Research Unit in Chivi Communal Area*, IIED, London

Mudhara, M, Chibudu, C, Mombeshora, B and Chikura, C (1996) 'A survey of farmers' experiences and perceptions of participatory resesearch in three communal areas of

Zimbabwe: Chivi, Murewa and Mutoko', pp13–26, in B Mombeshora, M Mudhara, S Chikura and Chibudu (compilers), *Generation of Appropriate Agricultural Technologies for the Smallholder Farming Sector*, proceedings of a workshop to review participatory research methodologies, Nyanga, May 13–16 1996, FSRU/DRSS, Harare, Zimbabwe

Mugwira, L M (1984) 'Relative effectiveness of fertilizer and communal area manures as plant nutrient sources', *Zimbabwe Agricultural Journal*, vol 81, pp81–89

Mugwira, L M and Mukurumbira, L M (1984) 'Comparative effectiveness of manures from the communal areas and commercial feedlots as plant nutrient sources', *Zimbabwe Agricultural Journal*, vol 816, pp241–250

Mugwira, L M and Mukurumbira, L M (1986) 'Nutrient supplying power of different groups of manure from the communal areas and commercial feedlots', *Zimbabwe Agricultural Journal*, vol 83, pp25–29

Mugwira, L M and Murwira, H K (1997) *Use of Cattle Manure to Improve Soil Fertility in Zimbabwe: Past and Current Research and Furture Research Needs*, CIMMYT, Harare

Mugwira, L M and Shumba, E M (1986) 'Rate of manure applied in some communal areas and the effect on plant growth and maize grain yields', *Zimbabwe Journal of Agricultural Research*, vol 83, pp99–104

Mukamuri, B and Murwira, H (1995) 'Local perceptions of soil fertility in Zimbabwe, a case study of Chivi, Mutoko and Mangwende communal area', in Proceedings of the Workshop on Nutrient Cycling and Soil Fertility Management in Africa, FARM-Africa, Ethiopia and IIED, London

Mukurumbira, L M (1985) 'Effects of rate of fertilizer nitrogen and previous grain legume crop on maize yields', *Zimbabwe Journal of Agricultural Research*, vol 826, pp177–179

Mulat, D (1996) 'Constraints to efficient and sustainable use of fertilisers in Ethiopia', pp242–266, in Mulat Demeke, Wolday Amha, Tesfaye Zegeye, Solomon Bellete and Simeon Ehui (eds), *Sustainable Intensification of Agriculture in Ethiopia*, Proceedings of the Second Conference of the Agricultural Economics Society of Ethiopia, Agricultural Economics Society of Ethiopia, Addis Ababa

Mulat, D et al (eds) (1996) *Sustainable Intensification of Agriculture in Ethiopia, Second Conference of Agricultural Economics Society of Ethiopia,* Addis Ababa

Mulat, D, Said, A and Jayne, T (1996) *Evaluating the Performance of Fertilizer Acquisition and Distribution in Ethiopia*, Ministry of Economic Development and Cooperation, Grain Market Research Project, Addis Ababa

Mulat, D et al (1997) *Promoting Fertilizer use in Ethiopia, The Implications of Improving Grain Market Performance, Input Market Efficiency and Farm Management*, MEDAC, Addis Ababa

Munguri, M W, Mariga, I K and Chivinge, O A (1996) 'The potential of optimizing cattle manure use with maize in Chinyika Resettlement Area, Zimbabwe', pp46–53, in *Research Results and Network Outputs in 1994 and 1995*, Proceedings of the Second Meeting of the Soil Fertility Network Working Group, CIMMYT, Harare

Murwira, H K (1994) 'Synchrony relationships of nitrogen release and plant uptake in a Zimbabwean soil amended with manure and fertiliser nitrogen', *African Crop Science Journal*, vol 2, pp69–77

Murwira, H K (1995) 'Ammonia losses from Zimbabwean cattle manure before and after incorporation into soil', *Tropical Agriculture*, vol 72, pp269–276

Murwira, H and Kirchmann, H (1993a) 'Carbon and nitrogen mineralization of cattle manures, subjected to different treatments, in Zimbabwean and Swedish soils', in K Mulongoy and R Merckx, *Soil Organic Matter Dynamics and Sustainability of Tropical Agriculture*, John Wiley and Sons, Chichester

Murwira, H K and Kirchmann, H (1993b) 'Nitrogen dynamics and maize growth in a Zimbabwean sandy soil under manure fertilisation', *Communication in Soil Science and Plant Analysis*, vol 24, pp2343–2359

Murwira, K H, Swift, M J et al (1995) 'Manure as a key resource in sustainable agriculture', in S Fernández-Rivera, T O Williams and C Renard, *Livestock and Sustainable Nutrient Cycling in Mixed Farming Systems of Sub-Saharan Africa,* J M Powell, ILCA, Addis Ababa, vol 2, pp131–148

Nicholson, S, et al (1998) 'Desertification, drought and surface vegetation: an example from the West African Sahel', *Bulletin of the American Meteorological Society*, vol 79, p5

Noordwijk, M van (1999) 'Nutrient cycling in ecosystems versus nutrient budgets of agricultural systems', pp1–26, in E Smaling, O Oenema and L Fresco (eds), *Nutrient Disequilibria in Agroecosystems: Concepts and Case Studies*, CABI, Wallingford

Nyamaphene, K (1991) *Soils of Zimbabwe*, Nehanda Publishers, Harare

Nyerges, A, E (1997) 'Introduction: the ecology of practice', in A E Nyerges (ed) *Introduction: The Ecology of Practice* Gordon and Breach Publishers, Newark, NJ

Oldeman, L R (1994) 'The global extent of soil degradation', in D J Greenland and I Szabolcs (eds) *Soil Resilience and Sustainable Land Use*, CABI, Wallingford

Oldeman, L R, Hakkeling, R T A et al (1990) *World Map of the Status of Human-Induced Soil Degradation: An Explanatory Note*, United Nations Environment Programme, Nairobi

O'Neill, R, DeAngelis, D et al (1986) *A Hierarchical Concept of Ecosystems*, Princeton University Press, Princeton

Östberg, W (1995) 'Land is coming up. The Burunge of central Tanzania and their environments', *Stockholm Series in Social Anthropology*, vol 34, Department of Social Antrhopology, University of Stockholm, Stockholm

PADLOS-OECD/CILSS (1998) *Decentralisation and Local Capacity Building in West Africa: Results of the PADLOS-Education Study*, CILSS, Ouagadougou

Palm, C A, Myers, R J K and Nandwa, S (1997) 'Combined use of organic and inorganic nutrient sources for soil fertility maintenance and replenishment', pp193–218, in R J Buresh, P A Sanchez and F Calhoun (eds), *Replenishing Soil Fertility in Africa*, SSSA, Wisconsin

Pankhurst, H (1993) *Vulnerability, Coping Mechanisms and The Food Security Project in Kindo Koisha, Welaita*, SOS-Sahel, Addis Ababa

Pausewang, S, Cheru, F, Brune, S and Chole, E (eds) (1990) *Ethiopia Rural Development Options*, Zed Press, London

Penning de Vries, F W T and Djiteye, M A (eds) (1982) 'La productivité des paturages sahélians, une étude des sols, des végétations et de l'exploitation de cette ressource naturelle', *Agricultural Research Reports* no 918, Pudoc, Wageningen

Pichot, J, Sedogo, M P, Poulain, J F and Arrivets, J (1981) 'Evolution de la fertilité d'un sol ferrugineux tropical sous l'influence de fumures minérales et organiques', *Agronomie Tropicale*, vol 36, pp122–133

Pieri, C (1989) *Fertilité des Terres de Savanes: Biland de Trente Ans de Recherche et de Développement Agricoles au Sud du Sahara,* Ministère de la Coopération et CIRAD-IRAT, Paris

Pieri, C (1995) 'Long-term soil management experiments in semi-arid francophone Africa', in Lal, R and Stewart, B A (eds), *Soil Management: Experimental Basis for Sustainability and Environmental Quality*, CRC Press, Florida

Pieri, C, Dumanski, J, Hamblin, A and Young, A (1995) 'Land quality indicators', *World Bank Discussion Paper*, no 315, World Bank, Washington DC

Piha, M I (1993) 'Optimizing fertilizer use and practical rainfall capture in a semi-arid environment with variable rainfall', *Experimental Agriculture*, vol 29, pp405–415

Pollet, E and Winter, G (1971) *La société Soninké*, Editions de l'Université' de Bruxelles, Brussels

Prudencio, C, Y (1993) 'Ring management of soils and crops in the west African semi-arid tropics: the case of the Mossi farming system in Burkino Faso', *Agriculture, Ecosystems and Environment*, vol 47, pp227–264

Ramisch, J (1999) 'In the balance? evaluating soil nutrient budgets for an agro-pastoral village of Southern Mali', *Managing Africa's Soils*, No, 9, IIED, London

Reij, C, Scoones, I and Toulmin, C (eds) (1996) *Sustaining the Soil: Indigenous Soil and Water Conservation in Africa*, Earthscan Publications, London

Richards, P (1985) *Indigenous Agricultural Revolution, Ecology and Food Production in West Africa*, Hutchinson, London

Richards, P (1989) 'Agriculture as a performance', pp39–42, in R Chambers, A Pacey, L-A Thrupp, (eds) *Farmer First, Farmer Innovation and Agricultural Research*, Intermediate Technology Publications, London

Rodel, M and Hopley, J (1969) *The Development of Small Intensive Farms at Henderson Research Station*, Henderson Research Station, Mazoe

Rodel, M G W and Hopley, J H D (1970) 'Investigations into systems of farming for tribal trust lands', *Rhodesia Zimbabwe Agricultural Journal*, vol 70, pp1–18

Rodel, M G W, Hopley, J D H et al (1980) 'Effects of applied nitrogen: kraal compost and maize stover on the yields of maize grown on poor granite soil ' *Zimbabwe Journal of Agricultural Research*, vol 77, pp229–232

Roe, E (1991) 'Development narratives, or making the best of blueprint development', *World Development*, vol 19, pp287–300

Rohrbach, D (1988) 'The growth of smallholder maize production in Zimbabwe, causes and implications for food security', PhD thesis, Department of Agricultural Economics, Michigan State University, East Lansing

Rohrbach, D (1989) 'The economics of smallholder maize production in Zimbabwe, implications for food security', *MSU International Development Paper*, no 11, Michigan State University, East Lansing

Röling, N and Jiggins, J (1997) 'The ecological knowledge system', in N Röling, A Wagemakers (eds), *Social Learning for Sustainable Agriculture*, Cambridge University Press, Cambridge, UK

Röling, N and Wagemakers, A (1997) *Social Learning for Sustainable Agriculture*, Cambridge University Press, Cambridge

Russell, E (1988) *Russell's Soil Conditions and Plant Growth*, Longmans, Harlow

Sanchez, P A and Logan, T J (1992) 'Myths and science about the chemistry and fertility of soils in the tropics', *Myths and Science of Soils of the Tropics: Special Publication no* 29, SSSA

Sanchez, P A, Shepherd, K D et al (1997) 'Soil fertility replenishment in Africa: an investment in natural resource capital', *Replenishing Soil Fertility in Africa*, pp1–46, in R J, Buresh, P A, Sanchez and F, Calhoun, SSSA, Wisconsin

Sanchez, P A, Izac, A-M N, Valencia, I and Pieri, C (1996) 'Soil fertility replenishment in Africa: a concept note', in S A Breth (ed), *Achieving Greater Impact from Research Investments in Africa*, Chitedze Research Station/Bunda College, Lilongwe, Malawi Sasakawa Africa Association, Mexico City

SARDC (1994) 'Soil erosion: southern African environmental issues', *CEP Factsheet*, no 1, SARDC, Harare

Sasakawa-Global 2000 (1995) *Agricultural Project in Ethiopia*, Sasakawa-Global 2000, Addis Ababa

Saunder, D H (1960) 'Soil productivity in relation to rotational and other cultural practices', Proceedings of the 5th Annual conference of the Professional Officers of the Department of Research and Specialist Services, Salisbury

Saunders, D (1959) 'What is the value of green manuring in Rhodesia?', Proceedings of the Fourth Annual Conference of Professional Officers of the Department of Research and Specialist Services, Ministry of Agriculture, Salisbury

Saunders, D and Grant, P (1962) 'Rate of mineralisation of organic matter in cultivated Rhodesian soils', *Transactions and Communications of the International Soil Science Society*, , vols IV and V, pp235–239

Scholes, R J (1990) 'The influence of soil fertility on the ecology of southern African dry savanas', *Journal of Biogeography*, vol 17, pp415–419

Scholes, R J, Dalal, R et al (1994) 'Soil physics and fertility: the effects of water, temperature and texture', in P L Woomer and M J Swift (eds) *The Biological Management of Tropical Soil Fertility*, John Wiley & Sons, Chichester

Schreyger, E (1984) *L'Office du Niger au Mali, la Problématique d'une Grande Entreprise Agricole dans la Zone du Sahel*, Steiner, Germany

Scoones, I (1991) 'Wetlands in drylands: key resources for agricultural and pastoral production in Africa', *Ambio*, vol 20, pp366–371

Scoones, I (1997) 'The dynamics of soil fertility change: historical perspectives on environmental transformation from Zimbabwe', *Geographical Journal*, vol 163, pp161–169

Scoones, I (1998) 'Investigating soil fertility in Africa: some reflections from research in Ethiopia and Zimbabwe', in L Bergström and H Kirchmann (eds) *Carbon and Nutrient Dynamics in Natural and Agricultural Tropical Ecosystems*, CABI, Wallingford, Oxon

Scoones, I, with Chibudu, C, Chikura, S, Jeranyama, P, Machaka, D, Machanja, W, Mavedzenge, B, Mombeshora, B, Mudhara, M, Mudziwo, C, Murimbarimba, F and Zirereza, B (1996) *Hazards and Opportunities: Farming Livelihoods in Dryland Africa: Lessons from Zimbabwe*, Zed Books, London

Scoones, I and Cousins, B (1994) 'The struggle for control over dambo resources in Zimbabwe', *Society and Natural Resources*, vol 7, pp579–594

Scoones, I and Thompson, J (eds) (1994) *Beyond Farmer First, Rural People's Knowledge: Agricultural Research and Extension Practice*, IT Publications, London

Scoones, I and Toulmin, C (1995) 'Socio-economic dimensions of nutrient cycling in agropastoral systems in dryland Africa keynote paper', pp353–370, in J M Powell, S Fernández-Rivera, T O Williams and C Renard, *Livestock and Sustainable Nutrient Cycling in Mixed Farming Systems of Sub-Saharan Africa*, ILCA, Addis Ababa, vol 2

Scoones, I and Toulmin, C (1998) 'Soil nutrient balances, what use for policy?' *Agriculture, Ecosystems and Environment*, vol 71, pp255–267

Scoones, I and Toulmin, C (1999) *Policies for Soil Fertility Management in Africa*, DFID, London

Scoones, I and Wolmer, W (eds) (forthcoming) *Pathways of Change: Crops, Livestock and Livelihoods in Africa – Insights from Ethiopia, Mali and Zimbabwe*, James Currey, Oxford

Scott, H (1950) 'Journey to the Gughe Highlands Southern Ethiopia, 1948–9; biogeographical research at high altitudes', Proceedings of the Linnean Society of London, London

Sebillote, M (ed) (1989) *Fertilité et systèmes de production*, INRA, Paris

Semb, G and Robinson, J B D (1969) 'The natural nitrogen flush in different arable soils and climates in east Africa', *East African Agricultural and Forestry Journal* January, pp350–370

Shantz, H and Marbut, C (1923) 'The vegetation and soils of Africa', *American Geographical Society Research Series*, no 13, New York

Shepherd, K D and Soule, M J (1998) 'Soil fertility management in west Kenya, dynamic simulation of productivity, profitability and sustainability at different resource endowment levels', *Agriculture, Ecosystems and Environment*, vol 71, pp131–146

Shepherd, K D, Ohlsson, E, Okalebo, J R, Ndufa, J K and David, S (1995) 'A static model of nutrient flow on mixed farms in the highlands of western Kenya to explore the possible impact of improved management', in J M Powell et al (eds), *Livestock and Sustainable Nutrient Cycling in Mixed Farming Systems of Sub-Saharan Africa*, ILCA, Addis Ababa

Shumba, E (1984) 'Reduced tillage in the communal areas', *Zimbabwe Agricultural Journal*, vol 81, pp235–239

Shumba, E (1985) 'Crop production under marginal communal area environments', *Zimbabwe Science News*, vol 19, pp39–43

Shumba, E (1986) 'On-farm research priorities resulting from a diagnosis of the farming systems in Mangwende, a high potential area in Zimbabwe', *Zimbabwe Agriculture Journal*, Special Report 5, DRSS, Harare

Shumba, E M, Dhilwayo, H H et al (1990) 'The potential of maize-cowpea intercropping in low rainfall areas of Zimbabwe', *Zimbabwe Journal of Agricultural Research*, vol 28, pp33–38

Sillitoe, P (1996) *A Place Against Time, Land and Environment in the Papua New Guinea Highlands*, Harwood Academic Publishers, Amsterdam

Smaling, E, Stoorvogel, J and Windmeijer, P (1993) 'Calculation of soil nutrient balances in Africa at different scales, II – district scale', *Fertiliser Research*, vol 35, pp237–250

Smaling, E (ed) (1998) 'Nutrient balances as indicators of productivity and sustainability in sub-Saharan African agriculture', *Agricultural Ecosystems and Environment*, vol 71, pp1–3

Smaling, E and Fresco, L O et al (1996) 'Classifying, monitoring and improving soil nutrient stocks and flows in African agriculture', *Ambio*, vol 258, pp492–496

Smaling, E, Oenema, O and Fresco, L (eds) (1997) *Nutrient Disequilibria in Agro-ecosystems: Concepts and Case Studies*, CABI, Wallingford

Solomon, B (1996) 'Sustainable intensification of agriculture, overview of concepts', pp1–11, in M Demeke, A Wolday, Z Tesfaye, B Solomon and E Simeon (eds), *Sustainable Intensification of Agriculture in Ethiopia*, Proceedings of the Second Conference of the Agricultural Economics Society of Ethiopia, Agricultural Economics Society of Ethiopia, Addis Ababa

SOS-Sahel (1993) *Koisha Rural Development Project, Income and Expenditure Survey Data*, SOS-Sahel, Bele

Ståhl, M (1974) *Ethiopia, Political Contradictions in Agricultural Development*, Scandinavian Institute of African Studies, Uppsala

Ståhl, M (1981) *New Seeds in Old Soil, A Study of the Land Reform Process in Western Wollega, Ethiopia 1975–76*, Scandanavian Institute of African Studies, Uppsala

Steiner, K G (1996) *Causes of Soil Degradation and Development Approaches to Sustainable Soil Management*, Margraf Verlag, Weikersheim

Stigand, C C H (1910) *To Abyssinia Through an Unknown Land: An Account of a Journey Through Unexplored Regions of British East Africa by Lake Rudolf to the Kingdom of Menelek*, Seeley and Co Ltd, London

Stocking, M (1987) 'Measuring land degradation', pp49–60, in P Blaikie and H Brookfield (eds), *Land Degradation and Society*, Methuen, London

Stocking, M (1994) *Six Myths About Soil Erosion: Or How to Fool Yourself and Others*, University of East Anglia, Norwich

Stocking, M and Elwel, H (1984) 'An assessment of soil erosion in Zimbabwe', *Zimbabwe Science News*, vol 19, pp27–33

Stoorvogel, J and Smaling, E (1990) *Assessment of Soil Nutrient Depletion in Sub-Saharan Africa*, Winand Staring Centre, Wageningen

Stoorvogel, J and Smaling, E et al (1993) 'Calculating soil nutrient balances in Africa at different scales: supra-national scale', *Fertilizer Research*, vol 35, pp227–235

Sumberg, J (1998) 'Mixed farming in Africa: the search for order, the search for sustainability', *Land Use Policy*, vol 15, pp293–318

Sutcliffe, J P (1993) *Economic Assessment of Land Degradation in the Ethiopian Highlands: A Case Study*, National Conservation Strategy Secretariat, Ministry of Planning and Economic Development, Addis Ababa

Swift, J (1996) 'Desertification, narratives, winners and losers', in M Leach and R Mearns (eds), *The Lie of the Land, Challenging Received Wisdom on the African Environment*, James Currey, London

Swift, M J (1998) 'Ten years of soil biology and fertility research: Where next?', in L Bergström and H Kirchmann, (eds) *Carbon and Nutrient Dynamics in Natural and Agricultural Tropical Ecosystems*, CABI, Wallingford

Swift, M J, Seward, P D et al (1994) 'Long-term experiments in Africa: developing a database for sustainable land use under global change', pp229–251, in R A Leigh and A E Johnston (eds) *Long-Term Experiments in Agricultural and Ecological Sciences*, CAB International

Swift, M, Frost, P, Campbell, B, Hatton, J and Wilson, K (1989) 'Nitrogen cycling in farming systems derived from savanna: Perspectives and challenges', pp63–76, in M Clarholm and L Bergstran (eds), *Ecology of Arable Land*, Kluwer, Amsterdam

Sy, O (1998) 'Interview', in *Newsletter of the Club du Sahel*, OECD, Paris

Syers, J K (1997) 'Managing soils for long-term productivity', *Phil Trans R Soc Lond*, vol 352, pp1011–1021

Takele, G (1996) 'Project in sustainable intensification of agriculture in Ethiopia', in *Sustainable Intensification of Agriculture in Ethiopia, Second Conference of Agricultural Economics Society of Ethiopia*, Addis Ababa

Talawar, S (1996) *Local Soil Classification and Management, Practices: Bibliographic Review*, Laboratory of Africultural and Natural Resource Anthropology, Athens, Georgia

Tempany, H (1949) 'The practice of soil conservation in the British Colonial Empire', *Technical Communication*, no 45, Commonwealth Bureau of Soil Science, Harpenden

Tempany, H, Rodder, G and Lord, L (1944) 'Soil erosion and soil conservation in the Colonial Empire', *Empire Journal of Experimental Agriculture*, vol 12, pp121–153

Teressa, A (1997) 'Factors influencing the adoption and intensity of use of fertiliser, the case of Lume district', Central Ethiopia, *Quarterly Journal of International Agriculture*, vol 36, pp173–187

Thompson, J and Purves, W (1981) 'A guide to the soils of Zimbabwe', *DRSS Technical Handbook*, no 3, DRSS, Harare

Tiffen, M, Mortimore, M and Gichuki, F (1994) *More People, Less Erosion, Environmental Recovery in Kenya*, John Wiley, Chichester

Toulmin, C (1992) *Cattle, Women and Wells, Managing Household Survival in the Sahel,* Clarendon Press, Oxford

Toulmin, C (1993) 'Combatting desertification, setting the agenda for a Global Convention, Dryland Networks Programme', *Issues Paper*, no 42, IIED, London

Toulmin, C (1997) 'The Desertification Convention', in F Dodds (ed), *The Way Forward: Beyond Agenda 21,* Earthscan, London

Toulmin C and Scoones, I (eds) (1998) *Final Report to EC/DGXII: Dynamics of Soil Fertility Management in Sub-Saharan Africa*, IIED/IDS, unpublished, London and Brighton

Trapnell, C (1943) *The Soils: Vegetation and Agriculture of North-Eastern Rhodesia*, Government Printer, Lusaka

Trapnell, C and Clothier, J (1937) *The Soils: Vegetation and Agriculture of North-Eastern* Rhodesia, Government Printer, Lusaka

Trounce, J M, Tanner, P D et al (1985) 'The effect of compound fertilizer, manure and nitrogen top dressing on maize grown on communal area vlei soils', *Zimbabwe Journal of Agricultural Research Zimbabwe*, vol 826, pp199–203

Truong, B (1998) 'Mise au point de formules d'engrais à partir des phosphates de Tilemsi', in H Breman and K Sissoko (eds), *L'intensification agricole au Sahel,* Karthala, Paris

Turner, B L, Hyden, G and Kates, R (eds) (1993) *Population Growth and Agricultural Change in Africa,* University Press of Florida, Gainseville

Turner, M (1995) 'The sustainability of rangeland to cropland nutrient transfer in semi-arid West Africa, ecological and social dimensions neglected in the debate', in J M Powell, S Fernández-Rivera, T O Williams and C Renand (eds), *Livestock and Sustainable Nutrient Cycling in Mixed Farming Systems of Sub-Saharan Africa, Volume II, Technical Papers*, International Livestock Centre for Africa, Addis Ababa

Tyson, P (1986) *Climate Change and Variability in Southern Africa*, Oxford University Press, Oxford

UNDP (1998) *Human Development Report*, Oxford University Press, Oxford

UNDP (1999) *Human Development Report*, United Nations Development Programme, New York

UNEP (1984) *General Assessment of Progress in the Implementation of the Plan of Action to Combat Desertification, 1978–1984*, UNEP, Nairobi

UNEP (1992) *World Atlas of Desertification*, Edward Arnold, London

UNEP (1994) 'General assessment of progress in the implementation of the plan of action to combat desertification, 1978–1984', Report of the Executive Director, UNEP/GC 12/9, UNEP, Nairobi

van der Fliert, E (1993) 'Integrated pest management: farmer field schools generate sustainable practices; a case study in central Java evaluating IPM training', *WU Papers* 93/3 Wageningen Agricultural University, The Netherlands

van den Bosch, H, Gitari, J, Ogaro, V, Maobe, S and Vlaming, J (1998) 'Monitoring nutrient flows and economic performance in African farming systems', *Agriculture, Ecosystems and Environment*, vol 71, pp63–80

van der Pol (1992) *Soil Mining, an Unseen Contributor to Farm Income in Southern Mali*, Royal Tropical Institute KIT, Amsterdam

van Duivenbooden, N (1992) *Sustainability in Terms of Nutrient Elements with Special Reference to West Africa*, Report 160, CABO-DLO, Wageningen

van Keulen, H and Breman, H (1990) 'Agricultural development in the West African Sahelian region, a cure against land hunger?' *Agriculture, Ecosystems and Environment*, vol 32, pp177–197

Van Veldhuizen, L, Waters-Bayer, A, Ramirez, R, Johnson, D and Thompson, J (eds) (1997) *Farmers' Research in Practice: Lessons from the Field*, IT Publications, London

Vanderheym, M J G (1896) 'Une expedition avec le negous menelik', *Le Tour du Monde*, no 9, 29 Feb 1896

Vincent, V and Thomas, R (1962) *An Agricultural Survey of Southern Rhodesia. Part 1, Agroecological Survey*, Government Printer, Salisbury

Vlaar, J C J (ed) (1992) *Les Techniques de Conservation des Eaux et des Sols dans les Pays du Sahel*, CIEH, Ouagadougou and UAW, Wageningen

Vogel, H (1992) 'Effects of conservation tillage on sheet erosion from sandy soils at two experiemental sites in Zimbabwe', *Applied Geography*, vol 12, pp229–242

Vogel, H (1994) 'An evaluation of five tillage systems for smallholder agriculture in Zimbabwe', *Der Tropenlandwirt*, vol 94, pp21–36

WADU (1972) 'WADU phase II proposal', prepared by Victor Burke, Project Director, Wolamo, Soddo, 28 November 1972, WADU, Soddo, Ethiopia

WADU (1973) *Credit Programme and Income Distribution*, WADU, Soddo, Ethiopia

WADU (1976a) *Agricultural Survey of Bolloso 1971*, WADU Wollaita Agricultural Development Unit, Soddo, Ethiopia

WADU (1976b) *Agricultural Survey of Bele, Wolayta 1971*, WADU Wollaita Agricultural Development Unit, Soddo, Ethiopia

WADU (1976c) *General Agricultural Survey of Wollaita*, WADU, Soddo

WADU (1977) *Summary of Crop Trial Results* (1971–1977, Crop and Pasture Section, Publication, no 55, WADU, Soddo, Ethiopia

WADU (1978) *WADU Project II Assessment*, WADU, Soddo, Ethiopia

Wales, M and Le Breton, R (1998) *Ethiopia Soil Fertility Initiative, Concept Paper*, FAO, Rome

Walker, B (1993) 'Rangeland ecology, understanding and managing change', *Ambio*, vol 22, pp80–87

Walling, D (1984) 'The sediment yield of African rivers', in *Challenges in African Hydrology and Water Resources*, IAHS Publication, vol 144, pp265–283

Wallis, J (ed) (1954) *The South African Diaries of Thomas Leask*, Chatto and Windus, London

Warren, G (1992) *Fertilizer Phosphorus Sorption and Residual Value in Tropical African Soils*, NRI, Chatham

Watson, J (1977) 'The use of mounds of the termite microtermes falciger germstacker as a soil amendment', *Journal of Soil Science*, vol 28, pp664–672

Watts Padwick, G (1983) 'Fifty years of experimental agriculture II: The maintenance of soil fertility in tropical Africa, a review' *Experimental Agriculture*, vol 19, pp293–310

Webb, P and von Braun, J (1994) *Famine and Food Security in Ethiopia, Lessons for Africa*, John Wiley and Sons, Chichester

Weigel, G (1986) 'The soils of Gununo area Kindo Koisha', *Soil Conservation Research Project SCRP, Research Report*, no 8, University of Berne, Berne

Wellby, M S (1901) *Twixt Sirdar and Menelik: An Account of a Year's Expedition from Zeila to Cairo Through Unknown Abyssinia*, Harper and Brothers Publishers, London

Westoby, M B, Walker et al (1989) 'Opportunistic management for rangelands not at equilibrium ' *Journal of Range Management*, vol 42, pp266–274

Westpahl, E (1975) *Agricultural Systems in Ethiopia*, Haile Selassie University, Addis Ababa and Agricultural University, Wageningen

Whitlow, R (1988) *Land Degradation in Zimbabwe: A Geographical Study*, University of Zimbabwe, Harare

Whitlow, R and Campbell, B (1989) 'Factors influencing erosion in Zimbabwe, a statistical analysis', *Journal of Environmental Management*, vol 29, pp17–30

Wilson, K (1989) 'Trees in fields in southern Zimbabwe', *Journal of Southern African Studies*, vol 15, pp1–15

Wilson, K (1990) 'Ecological dynamics and human welfare, a case study of population, health and nutrition in Zimbabwe', DPhil thesis, University of London, London

Winrock International (1992) *Assessment of Animal Agriculture in Sub-Saharan Africa*, Winrock International, Morrilton

Wolmer, W and Scoones, I (2000) 'The science of 'civilised' agriculture, the mixed farming discourse in Zimbabwe', *African Affairs*, vol 99, pp575–600

Wood, A and Ståhl, M (1998) *Ethiopia National Conservation Strategy, Phase One Report*, International Union for the Conservation of Nature, Gland

Woomer, P L and Muchena, F N (1996) 'Recognizing and overcoming soil constraints to crop production in tropical Africa', *African Crop Science Journal*, vol 4, pp503–518

Woomer, P L, Martin, A et al (1994) 'The importance and management of soil organic matter in the tropics', pp47–80, in P L Woomer and M J Swift (eds), *The Biological Management of Tropical Soil Fertility*, John Wiley & Sons, Chichester

Woomer, P L and Swift, M J (eds) (1994) *The Biological Management of Tropical Soil Fertility*, Wiley, Chichester

Woomer, P L, Martin, A, Albrecht, A, Resck, D and Scherpenseel, H (1994) 'The importance and management of soil organic matter in the tropics', in P L Woomer and M J Swift (eds), *The Biological Management of Tropical Soil Fertility*, Wiley, Chichester

Worku, T (2000) 'Stakeholder participation in policy processes in Ethiopia', *Managing African Soils*, vol 17, IIED, London

World Bank-FAO (1996) *Recapitalisation of Soil Productivity in Sub-Saharan Africa*, World Bank, Washington DC/FAO, Rome

Wortmann, C and Kaizzi, C (1998) 'Nutrient balances and expected effects of alternative practices in farming systems of Uganda', *Agriculture, Ecosystems and Environment*, vol 71, pp115–130

Young, A (1980) *Tropical Soils and Soil Survey*, Cambridge University Press, Cambridge

INDEX

Page references in *italics* refer to tables, figures and boxes